# THE
# DEVIL'S ELEMENT

# THE
# DEVIL'S ELEMENT

PHOSPHORUS

AND

A WORLD OUT OF BALANCE

## *Dan Egan*

**W. W. NORTON & COMPANY**
*Celebrating a Century of Independent Publishing*

For information about permission to reproduce selections from this book, write to
Permissions, W. W. Norton & Company, Inc., 500 Fifth Avenue, New York, NY 10110

For information about special discounts for bulk purchases, please contact
W. W. Norton Special Sales at specialsales@wwnorton.com or 800-233-4830

Manufacturing by Lakeside Book Company
Book design by Brooke Koven
Production manager: Anna Oler

ISBN 978-1-324-00266-6

W. W. Norton & Company, Inc., 500 Fifth Avenue, New York, N.Y. 10110
www.wwnorton.com

W. W. Norton & Company Ltd., 15 Carlisle Street, London W1D 3BS

1 2 3 4 5 6 7 8 9 0

*In memory of Christopher Marsh*

# CONTENTS

# NOTE TO READERS

PURE PHOSPHORUS, element fifteen on the periodic table, does not typically exist on its own in the natural world. Even though phosphorus atoms are required by every living cell on Earth, they are naturally bound up with four oxygen atoms to form molecules called phosphates. This isn't a book about oxygen atoms, so in most all cases I will use the term phosphorus, though there will be cases in which I quote people using the term phosphate.

Similarly, I am aware that the preferred scientific term for an explosion of a type of nuisance (and sometimes poisonous) phytoplankton is an *algal* bloom. But the general public typically uses the term algae bloom, and the general public is the target audience for this book.

Along those lines, this book is not intended to be the last word on phosphorus. Many in the general public might not yet be aware of the phosphorus troubles the world is headed toward because of the element's dual roles as a dangerously potent toxic algae booster and as an essential—and increasingly scarce—

crop nutrient. But there are scientists who have been on these problems for years, and there are a host of emerging technologies and practices to address one or both sides of the overuse-scarcity problem tied to phosphorus. This book does not provide a survey of them. I discuss some potential paths forward to put the world back into a better phosphorus balance, but this book isn't intended to provide a prescription for the phosphorus paradox. It is an introduction to it.

# INTRODUCTION

IN LATE SUMMER 2018, Abraham Duarte was roaring down a neighborhood street in Cape Coral, Florida, at highway speed when a constellation of blue and red lights started to twinkle in his rearview mirror. He ditched his black Lexus and tried to escape on foot into a tangle of backyards, but he ran out of lawn sooner than expected.

As the hard-breathing cops closed in, bodycams rolling, Duarte had two choices. He could turn around, face the jangling handcuffs and the consequences of resisting arrest after being pulled over for speeding. Or the twenty-two-year-old, who sports a tattoo on his left arm that reads "Taking Chances," could swim for it.

Duarte plunged into one of the canals that cut through Cape Coral neighborhoods like so many nautical alleyways. It was the wrong choice. The problem wasn't that Duarte didn't know how to swim. It was that he didn't exactly jump into water; the surface of the canal was smothered in a brilliantly green algae goop thick as oatmeal—and poisonous.

"Help, I need help! I'm going to die!!" Duarte screeched as toxic fumes overwhelmed him while officers on the canal bank anxiously radioed for backups to bring a rescue rope. Duarte's face started to bob into the slime. One officer coached him to float on his back—to keep his mouth and nose above the poison.

"Fuck dawg. Fuck. Fuck. Damn. Holy shit!!" Duarte wailed as he tried to dog-paddle to shore.

"You need to get out of that stuff, that is gonna mess you up!" shouted one officer. "Seriously man, that is going to kill you."

Duarte's feet finally hit bottom when he got within several yards of the canal bank. At that point it was clear he wasn't going to drown, but he wasn't out of trouble. He began to vomit violently.

As Duarte neared the canal wall, he reached up and the rubber-gloved cops heaved him ashore. They rolled him onto his stomach, handcuffed him and gave him a quick shower with a garden hose that was nearly as green as the muck coating his eyes, nostrils, and throat. Muck that Duarte said smelled like "human feces." He was taken to the hospital and later charged with resisting arrest (without violence) and possession of a controlled substance.

Days after Duarte's close call in the canal, even as he was recuperating from being treated for gastrointestinal and respiratory distress, television news anchors across the country smirked as they narrated the cops' body camera chase video. The clip might have added to the legend of "Florida Man"—the internet's fixation on Sunshine State males who make headlines for committing bafflingly stupid acts.

But Duarte's plunge into that toxic quagmire is more than a meme.

It is an omen.

◊　◊　◊

THAT SAME SUMMER in the little city of Stuart, a beach community on the opposite side of the Florida peninsula from Cape Coral, some hundred panicked homeowners showed up at city hall in the middle of the business day to demand something be done about the same green goo plaguing their own coastal waters. It was a sweltering July day, the kind towns like Stuart are built for, but signs on the boardwalk outside city hall warned visitors:

BLUE-GREEN ALGAE—AVOID CONTACT WITH THE WATER

As people at the meeting introduced themselves and stated their affiliations, it became clear this was not a typical gathering of environmentalists. They weren't strategizing about how to protect some beleaguered species and the far-away lands or waters upon which it depends. These people, who represented businesses as well as homeowners' associations and fishing and yachting clubs, spoke as though *they* were the threatened species.

"I need help," said Will Embrey, a scraggly commercial fisherman whose business had collapsed right along with the region's schools of mackerel not long after the green slime arrived. "There are a lot of people like me that need help." The forty-five-year-old was suffering chronic stomach pain that was initially diagnosed as diverticulitis, and then ulcerative colitis, and then Crohn's disease. Eventually doctors had given up trying to figure out what made Embrey so sick.

Embrey didn't need to spend tens of thousands more dollars on more specialists, CT scans, and lab tests to figure out the nature of his illness. He knew it was the poisoned water, and he wasn't alone.

So many patients had been showing up in local clinics and emergency rooms reporting mysterious respiratory issues and gastrointestinal illnesses that in the days before the city hall meeting the head of the local health network had declared a public health "crisis." To gauge the scope of the algae scourge that had become a summer fixture in Stuart over the previous few years, he instructed clinics to begin asking these patients whether they had been swimming or had other contact with open waters—not good news for a place billed as one of Florida's top ten beach towns.

"Un-be-lievable, ladies and gentlemen," said the host of the meeting, a local Republican politician. "For anybody out there listening—this is real! This is happening!"

A FAMOUS ECOLOGIST named John Vallentyne made a dire prediction in the early 1970s: Decades of reckless industrial and municipal polluting threatened to do to North America's rivers and lakes something akin to what reckless farming had done to the Great Plains during the Great Depression—when drought-parched, wind-whipped soils seeded "black blizzards" fierce enough to blind jackrabbits and turn hundreds of thousands of prairie settlers into environmental refugees.

"Sometime before the year 2000, unless something is done to avert the situation," said Vallentyne, who was the chief scientist for Canada's Department of Fisheries and Oceans, "we shall find ourselves in the middle of an Algal Bowl, with effects on water comparable to those on land during the great American Dust Bowl of the 1930s."

In some nearshore areas, Lake Erie's famously walleye-rich waters had by the 1960s already degraded into a soupy green fishless stew. The type of algae smothering Erie and other water bodies at the time wasn't poisonous, but it was lethal nonetheless. The green mats, which sometimes spanned hundreds of square miles, sucked so much oxygen out of the water when they decomposed that they created vast "dead zones," in which almost nothing could survive. By the beginning of the 1970s, Lake Erie had become known as America's "Dead Sea."

Lawmakers spanning the political spectrum responded by enacting the Clean Water Act, which brought sweeping water quality improvements to Lake Erie and waters across the continent by finally prohibiting industries and cities from treating public waterbodies as waste lagoons.

To speed along the cleansing and prod future generations to do better, Vallentyne took on a second job—and personality. He began appearing as "Johnny Biosphere" at auditoriums across North America dressed in safari garb and with a beach-ball-size globe strapped to his back. His message: "Be kind to the Earth, and it will be kind to you!"

Johnny Biosphere often spoke to audiences of eight-year-olds, so his slogan stopped short of the obvious, opposite, and ominous corollary—abuse the Earth, and the Earth will abuse you.

This is not a children's book.

ONE OF THE signature elements of the Clean Water Act is that any industry or municipality that drips even a drop of pollution into public waters must first apply for a permit to do that polluting. The idea is that those permits must be regularly renewed, and as waste treatment technology improves, the amount of waste a polluter is authorized to discharge must steadily ratchet down.

The law worked brilliantly, but it left a yawning exemption for one specific industry—agriculture. There was a legitimate rationale for the loophole. It's relatively easy to filter what gurgles out of a pipe or scrub what floats out of a smokestack. You can't exactly use a broom brush to sweep a farm field of its excess pesticides and fertilizers that pollute public waters when they wash away in rains.

Of course, you *can* regulate what and how much fertilizer farmers put on their fields in the first place, but lawmakers nevertheless opted to essentially give them a pass.

This failure of the Clean Water Act to adequately police the agriculture industry is at the root of today's wave of algae infestations because the fertilizer washing off agricultural lands is a primary driver of the algae blooms. Worse, much of the green goo smothering lakes and ponds across the United States today isn't actually algae but primitive forms of photosynthesizing bacteria that collectively produce a suite of toxins, some powerful enough to rival anything cooked up in a military lab. One type of algae toxin is, in fact, so potent scientists gave it a punk rock name—"Very Fast Death Factor."

If you haven't heard about the emerging menace of blue-green algae that produce poisons known as cyanotoxins, you will.

In 2021 alone, media outlets across the country reported some four hundred US bodies of water were infested with the green slime, a 25 percent jump over the previous year. Outbreaks plague beaches from Biloxi, Mississippi, to Lewiston, Maine, to Madison, Wisconsin, to Spokane, Washington. Between 2017 and 2019, more than three hundred people across the United States ended up in emergency rooms after being exposed to toxic algae-contaminated waters. One bloom on Lake Erie in 2014 contaminated the water supply for some five hundred thousand people in Toledo, Ohio.

As of this writing, the only human deaths officially confirmed to have been caused by blue-green algae occurred in Brazil in the late 1990s after an outbreak in a public water supply killed dozens at a dialysis center, but there are worrying signs that more could come. They may already have.

In 2002, a Wisconsin coroner pointed to a cyanotoxin in the mysterious death of a seventeen-year-old boy who had hopped a fence and jumped into a blue-green algae-ravaged golf course pond to cool off on a hot July night.

An algae toxin was also a prime suspect in the perplexing case of a young California family found dead on a hiking trail along the blue-green algae-infested Merced River near Yosemite National Park in late summer 2021. Water samples from the river along the trail showed concerning levels of Very Fast Death Factor (anatoxin-a), though authorities later settled on hyperthermia as the culprit.

These toxic algae blooms, which regularly kill pets that only so much as splash in contaminated water, are not contained to this continent. And they threaten more than dogs; in 2020 the government of Botswana blamed a toxic bloom for killing 356 African elephants that drank from contaminated puddles.

Blue-green algae, also known as cyanobacteria, have been around for billions of years, but climate change is boosting the environmental role they play because they thrive in our warming waters and feast on the ever-increasing levels of atmospheric carbon.

Another cyanobacteria booster: fingernail-sized invasive zebra and quagga mussels spreading, cancer-like, across North America. Waters infested by the Caspian Sea basin natives are particularly vulnerable to toxic algae outbreaks because the filter-feeding mollusks will eat almost anything floating in the water *except* blue-green algae. This gives the blue-green algae a competitive edge over the nontoxic species of algae that make up the base of a healthy lake's food chain. This means when a mussel-infested lake suffers an algae outbreak, it is more likely to be of the toxic sort.

But arguably the most important driver in the surging blue-green algae blooms is one many people never think about. To begin to understand how this element in the toxic algae equation threatens waters worldwide, you need only to travel a little more than a hundred miles northwest of Stuart, Florida. There lies the essential root of Florida's blue-green algae problems, as well as similar water troubles across the continent—even though nobody at the Stuart City Hall meeting back in 2018 seemed to have the slightest notion of how a relatively desolate sweep of central Florida, just a couple of hours' drive to their northwest, could have anything to do with their public health crisis. The place is called Bone Valley.

ABOUT THIRTY-FIVE MILES east of Tampa is a quirky tourist attraction whose signature feature is a crane shovel big enough to scoop up several dump trucks' worth of rocks and stones. Little kids play, sandbox-style, in the tongue of pebbles spilling from its mouth. Bigger kids (and their parents) sift the pile for traces of long-gone beasts.

Because the sandy finger of land that is present-day Florida has for millions of years bobbed above and below the waterline as ocean levels have surged and shrunk, the center of the peninsula today is rich in fossilized remains from both land and sea. So rich that in the 1980s the little city of Mulberry converted a couple of old railroad cars into something of a fossil museum.

The museum is in the heart of Florida's Bone Valley region that sprawls across more than a million acres in the west-central part of the state. Here the fossilized remains of gargantuan armadillos lie buried among claws of extinct ground sloths that stood more than twelve feet tall. Remnants of elephant-sized mastodons are in the mix, as are whales, sea turtles, and megalodons—a long-gone species of jumbo shark with a mouth big enough to swallow a car.

The late 1800s discovery of this oddly matched cast of pre-historic characters frozen in time gripped the imagination of a public still coming to terms with the implications of Darwin's theory of evolution.

"In this vast antediluvian sepulcher one may give free rein to the imagination, and in weird fancy resurrect the strange forms of animal life that walked the earth when this beautiful peninsula was but a straggling line of sand dunes and coral reefs," reported one newspaper in 1890.

But central Florida's troves of prehistoric remains, the author noted, weren't just valuable as museum pieces. "To the practical man, the man with a purpose, the wealth seeker and the capitalist, these prodigious accumulations of [fossils] are Fortune's offerings—the opportunity of a lifetime."

He even predicted the fossils would become more precious to Florida than gold was to California in the 1850s. The stony remains of so many long-gone creatures (and, more importantly, so much of the sedimentary slabs of rock and pebbles entombing them) would actually prove to be far more valuable—because you can't sprinkle gold on a crop to grow food.

Florida's fossil beds and their surrounding sedimentary rock could, it turned out, be pulverized and soaked in acid to create a staggeringly potent fertilizer that spurs agricultural crops to grow at incredible rates. Twenty-seven of these fertilizer mines sprawl across nearly a half-million acres of central Florida. Nine of the mines remain active today, and for every ton of crucial nutrients miners pull from the ground another five tons of a mildly radioactive waste material is produced—waste that is piled into mini mountains in Florida's interior. These mounds of pollution are out of sight and mind for most Floridians—except when their waste periodically seeps off site and threatens the state's groundwater supplies as well as its coastal waters.

Yet the toxic piles of mine waste are allowed to keep growing because the fertilizer rock deposits at Bone Valley, along with a relatively small number of similar deposits scattered around the globe, are a big reason Earth's food production has been able to double in the last half century—right along with its population.

These rocks are why today's corn stalks, descendants of a wispy, grain-rich type of grass that Native Americans first cultivated nearly ten thousand years ago, grow to apple-tree heights and why bushel-per-acre yields have exploded almost five-fold since agriculture pioneers unleashed the powers of rock fertilizer on corn and other crops.

But there is a dark side to this chemical fertilizer's miraculous ability to make things grow—its potency doesn't fade when it hits water. And much of today's rock-based fertilizer spread by farmers gets washed off croplands before it can be taken up by plant roots. So instead of making bumper crops of food, it tumbles into our streams, rivers, and lakes, where it then fertilizes bumper crops of blue-green algae.

Nobody pondered the downside of tinkering with nature in this fashion at the time the Bone Valley rock fertilizer deposits were discovered in the late nineteenth century.

Floridians were so dazzled by the jackpot under their boots— about 75 percent of the rock fertilizer consumed in the United States today still comes from Florida—that newspapers of the time wrote stories of men willing to shoot each other over roadbed pebbles that had been laid down before they were recognized as nutritional gold.

But what, precisely, makes these rocks so precious?

Phosphorus.

PHOSPHORUS IS ESSENTIAL to plant growth, and that makes it essential to us, but the element is important beyond helping to grow our food. Phosphorus helps turn the meals we eat into the chemical energy that moves our muscles. Phosphorus is also crucial to our physical structure, in the biggest ways and in the smallest. Our bones and teeth are made with phosphorus. Phosphorus is also in our DNA. In fact, it *is* our DNA. The rails of the famous twisting helices that form the genetic blueprints that bring to life every single cell on the planet are made of phosphorus. From the corn we grow, to the animals that eat it, to the people who eat those animals, phosphorus is critical every step of the way.

No phosphorus, no life on Earth.

The same thing, of course, might be said about any number of life-essential elements, including the two other key components of modern fertilizer—nitrogen and potassium.

But there is a key difference between phosphorus and those other life-sustaining elements. The Earth still has loads of potassium reserves in the form of deposits left over from ancient dried-up sea beds and we are in no danger of running out of the stuff anytime soon. As for nitrogen, it is the most abundant gas in the atmosphere, and technology has existed since the early 1900s to pull it out of air and put it into a form suitable to be spread on farm fields. This means there is little worry about a looming global scarcity of these two elements to make the fertilizer that makes our crops grow big and fast enough to feed the planet's ever-ballooning population.

Phosphorus is a whole different story.

The initial source of the element that brought Earth to life is the magma that hardened into rock as early Earth cooled. Eventually wind and waves unleashed the traces of phosphorus in those igneous rocks. Once set loose, phosphorus atoms whisked and winged between the living and dead. When animals defecated or died and decayed, the phosphorus in their wastes and carcasses was absorbed by plants. When those plants died or were eaten and expelled, they released that same phosphorus to fuel a new generation of greenery that, in turn, sustained the next generation of grazing animals—and the people feasting upon them. And so on.

Phosphorus is the elemental link that completes the circle of life. Literally nothing else can do its job.

"We may be able to substitute nuclear power for coal, and plastics for wood, and yeast for meat, and friendliness for isolation—but for phosphorus there is neither substitute nor replacement," famed scientist and author Isaac Asimov observed in 1959.

It wasn't until the nineteenth century that humans discovered they could remove the phosphorus cap on plant growth—and Earth's population—by going straight to rare caches of phosphorus-rich sedimentary rock scattered around the globe. These deposits were created by dead organisms falling to the ocean floor that, over the course of millions of years, piled upon each other, like snowflakes on a glacier, to create so much mass and pressure that they fused into phosphorus-rich sedimentary rock. Geological forces eventually heaved some of this rock to Earth's surface, and these are the mineable deposits that now allow us to harvest lodes of phosphorus in a year that previously took eons to leach into the living world.

Humans might have figured out how to crack Asimov's phosphorus bottleneck with this rock mining, but the element today is precious and finite in the same way fossil fuels are. Yet we are blowing through Earth's accessible deposits at such a pace that, just like oil production, some scientists now fear we could hit "peak phosphorus" in just a matter of decades, at which point we risk declining mining yields—and chronic food scarcity.

"This is the gravest natural resource shortage you've never heard of," an editorial in *Foreign Policy* magazine proclaimed more than a decade ago.

The prospect of phosphorus shortages has only grown since, exacerbated by the wasteful way we are managing the dwindling reserves of a substance that is both a precious resource and a nasty pollutant. Globally, annual phosphorus rock harvests have roughly quadrupled since Lake Erie's Dead-Sea days a half century ago. Yet so much of the phosphorus we mine today and spread as fertilizer gets flushed off farm fields before it ever gets picked up by crops, let alone livestock, not to mention us. And much of that phosphorus that does make its way to the food on our dinner tables then makes its way, via sewer lines, into our waters instead of back onto croplands. Call it the phosphorus paradox—at the same time as we are drawing down our increasingly precious caches of mineable phosphorus rock we are overdosing our waters with it.

Some have predicted existing phosphorus reserves will play out by the end of the century, a time frame scoffed at by many who are knowledgeable about the issue, including those in the fertilizer business. But whatever the number of years, it is undeniable that we have cracked the circle of life and turned it into a straight line, and that line has an end, whether it's in one hundred years or four hundred years. The trouble won't come when the last of Earth's phosphorus-rich rock reserves have been mined, milled, and spilled into our waters. It will come instead when phosphorus deposits play out in certain regions of the world, leaving relatively few countries—even a few people—largely in control of the fertilizer gusher that sustains seven billion souls, and that day could be coming faster than you think.

Florida miners are on pace to run out of available rock in as few as thirty years, at which point the United States is at risk of becoming dependent on other countries to sustain its agriculture system.

Whether those countries share an interest in maintaining our nutritional security is another question. Roughly 70 to 80 percent of the globe's remaining phosphorus reserves are located in Morocco and the Western Sahara territory that Morocco has occupied—sometimes violently—since the 1970s. For one country, essentially one guy—the king of Morocco—to control so much of something every soul on the planet so desperately needs is a recipe for global instability, or worse.

THE SEVENTEENTH-CENTURY ALCHEMIST who discovered pure phosphorus by cooking it from his excrement (it is, after all, in every one of our cells) knew he had unleashed something uncanny—waxy white nuggets that smelled faintly of garlic and cast a bewitching glow. He named his discovery after the Greek word for planet Venus—*phosphoros*. The word roughly translates into "Bringer of Light." It is a good name for such a glimmering element because Venus's twinkle in the predawn sky heralds the imminent arrival of the sun.

The Latin word for Venus translates similarly: *luc* (light) *fer* (bearer). Lucifer.

That actually would have been a better name for the alchemist's discovery because he soon learned that his curious nuggets had a propensity to spontaneously combust and burn as ferociously as anything that dripped from the nib of Dante's quill.

It wasn't long, in fact, before people started referring to phosphorus as the Devil's Element, and not only because it happened to be the thirteenth element discovered. The name stuck because of its dastardly toxicity (it is still an active ingredient in rat poison) and its explosiveness (as I write this phosphorus bombs are reportedly being used—probably illegally—by Russian forces occupying Ukraine).

The idea of phosphorus as the Devil's Element is, in fact, even more appropriate now.

The biblical devil coaxed Earth's first human inhabitants to nibble on an apple from the tree of knowledge, an awakening that, as the story goes, exiled Adam and Eve from their paradise and forced them to scavenge what they could from an unforgiving planet.

Their struggle remains ours today. For the last century we've been winning it as we awakened to the power of mined phosphorus fertilizer and all the fruits it bears. But hitching our existence to mined phosphorus in this fashion carries its own Faustian burden. In exchange for breaking the natural throttle that limited how many humans Earth could sustain, we are polluting freshwaters with phosphorus fertilizer to the point those waters are increasingly prone to be too fouled to swim in, to fish upon, and to drink from. We are soiling our own garden.

The only thing we can do now to protect and restore those waters and at the same time ensure there is enough phosphorus—enough food—available for all the souls yet to be born is to train this latter-day devil to chase its own tail, to restore the virtuous phosphorus circle of life that we've broken.

This will require a dramatic change in how much chemical fertilizer we use and how we use it. It's also going to take a revolution in the way we manage civilization's waste streams, human waste included.

The costs for not taming the Devil's Element in this manner are already starting to mount.

The same week Abraham Duarte took his fateful plunge into the Florida canal, a local newspaper reported that the sale of a $7 million waterfront home had fallen through because of toxic algae. The real estate agent handling the deal sounded almost as panicked as Duarte had been in the canal. If Florida's water goes bad, he said, so does . . . everything.

"The environment," he said, "is all Florida has to offer."

The same is true for Earth itself.

*Part I*

# THE RACE
# FOR
# PHOSPHORUS

# CHAPTER 1

## *The Devil's Element*

ALITTLE OVER a decade ago, Gerd Simanski, a retired German department store manager, picked up an unexpected hobby after he and his wife purchased a tidy brick vacation home in a tiny village not far from the Baltic Sea: beach-combing. Simanski especially liked hunting for the fossilized remains of belemnites, squid-like predators that jetted through Jurassic seas by sucking in water and then blasting it out a tube near their mouth in a manner that sent them darting backwards.

Simanski liked scouring the beach for nuggets of amber and fossilized sea creatures because it somehow buoyed his mood to ponder how insignificant humans are on the stage of planetary history. When he and his wife were first shopping for a retire-ment property, he was often told how new-construction homes have a "lifetime warranty"—of thirty years. Simanski's bushy mustache bends up and his eyes crease from his wry smile when he tells me in the next breath that the belemnite fossil he just put in my hand has been around for tens of millions of years.

He found beachcombing so relaxing that he would go out

alone for hours on end in all weather, even on the drizzly and frigid morning of January 13, 2014, when he put on his winter jacket, grabbed his car keys and told his wife he would be back for lunch in a couple of hours.

Simanski was alone on the shoreline that day as he ambled, eyes down, over the sweep of rocks squeezed between the tongues of the lazy, glassy Baltic waves and a thirty-foot-high bluff when he spotted what he thought was a piece of fossilized oyster shell about the size of a US quarter. He didn't consider the orangish stone a prize find, but he thought it was worth bringing home to show his wife. So the sixty-eight-year-old stooped over, picked it up and dropped it in his pants pocket. Then he moved on, looking for something a little more interesting.

After about ten minutes, Simanski heard a pop and felt a sharp pain near his hip, and when he looked down he saw yellow flames flaring from his left leg. "It was like lightning coming out of my jeans. Like a flash," said Simanski, who was initially more baffled than frightened. "It's cold. It's raining. It's wet and I'm thinking, where did this flash come from? I don't smoke. I don't have a lighter. It just could not be."

Bewilderment turned to terror after Simanski shoved his hand into his pocket to snuff out whatever it was that caused the fire and he felt nothing but a viscous substance that had the consistency of melted chocolate. When he jerked his hand out of his pocket each fingertip was covered in the goo and ablaze like its own candle.

Simanski screamed for help as he watched the flames that had scorched the skin off his thigh begin to cook through the pale yellow fat beneath "like sizzling bacon." He yelled toward a lone fisherman down the beach to call an ambulance and then he instinctively headed for the water. The flames flickered out as he splashed into the frigid sea and Simanski, afraid the fire would roar back if he stepped ashore, remained in the surf for

4

nearly a half hour waiting for help, shivering, scorched, and in shock.

When two police officers finally arrived and coaxed him ashore, they saw flesh as blackened as an over-grilled chicken leg, a sight so ghastly both officers later had to take leaves of absence from work. There was talk about calling in a medical helicopter, but it wasn't dispatched out of fears the mysterious flames would reappear midflight and take down the aircraft. The paramedics who finally arrived by ambulance cut off what was left of Simanski's jeans, wrapped him in a blanket and zoomed off toward the emergency room. They, too, were so rattled by all the seared flesh that they couldn't find a vein to administer morphine as the ambulance screamed down rolling roads narrower than a suburban American driveway.

Simanski would spend most of the next two months in the hospital mending from burns to one-third of his body. Today he is largely healed but remains in chronic pain and needs pills to sleep. The damage to his left leg was so severe that the grafted skin holding it together is rutted and rough as tree bark.

He still can't fathom all that happened to him after he picked up that cold, wet stone. "It was just a rock," he says. "A little rock. A very little rock."

This was not an isolated incident, and the exploding little pebbles Simanski and other Baltic beachcombers have been finding in recent years are neither rock nor fossil. Many of the golden or orange nuggets plucked from the beach and the banks of the nearby Elbe River bear an uncanny resemblance to amber—fossilized pieces of tree resin for which the Baltic region is famous. But they aren't gemstones. They are actually fragments of some of the most dangerous stuff you can find on the periodic table—elemental phosphorus.

These nubs of pure phosphorus don't belong in the natural world any more than, say, a Styrofoam cup. That's because phosphorus atoms in their natural state are bonded with oxygen atoms to create various compounds known as phosphates—molecules that are essential to every living thing on Earth. Phosphates are a critical component of DNA. They help fuel the chemical reactions that release energy on the cellular level. They are the building blocks of cell walls and membranes, and they play an essential role in converting sunshine into Earth's greenery. Phosphates, simply put, bring life to a planet that would otherwise be a cold, dead rock.

But when phosphorus atoms do somehow shed their bonds with oxygen atoms, it is often only a temporary situation, one that typically ends explosively. All it takes for a nugget of pure phosphorus to burst into flames is for it to warm just a little above room temperature.

Elemental phosphorus nuggets are, in fact, so unnatural that for every one of them surfacing on beaches and riverbanks across northern Germany in recent years there is a story behind it, a story that has human fingerprints all over it.

Understanding how these pebbles got there requires a step back in time—a little more than seven decades, to be specific.

ON JULY 21, 1943, Hans Nossack, a coffee merchant and part-time writer, left his home in Hamburg for a two-week vacation—from work, and from the war that had been raging for the previous four years. The cottage he rented was a good ten miles from Hamburg city limits, but three nights after arriving the couple was roused from their slumber by an air raid siren warning of an attack on the city. "I jumped out of bed and ran barefoot out of the house, into this sound that hovered like an oppressive weight between the clear constellations and the dark earth, not here and not there but everywhere in space; there was no escaping it . . ." Nossack recalled just weeks later. "It was the sound of eighteen hundred airplanes approaching Hamburg from the south at an unimaginable height."

The plan to unleash the roaring swarm of bombers on Germany's northern industrial hub was seeded by Prime Minister Winston Churchill and President Franklin Roosevelt at a secret meeting earlier in 1943 in North Africa. They directed their military leaders to basically hold nothing back in the future aerial bombardment of German cities. The first objective in the one-page *Casablanca Directive*: "The progressive destruction and dislocation of German military, industrial and economic systems, and the undermining of morale of the German people to a point where their capacity for armed resistance is fatally weakened."

It might have been more honest to swap out the word "morale" in the directive for "lives," because the bombs of that era dropped on cities thousands of feet below were anything but precise. "We believe the Nazis and Fascists have asked for it," Roosevelt explained to Congress. "And they are going to get it."

The British were even more graphic in their public statements of what they intended to do to the German population. "The Nazis entered this war under the rather childish delusion that they were going to bomb everybody else and nobody was going to bomb them," Sir Arthur "Bomber" Harris, head of the British Royal Air Force, proclaimed. "At Rotterdam, London, Warsaw, and half a hundred other places, they put that rather naive theory into operation." Then Harris used an Old Testament phrase to put fear in the minds of German civilians. "They sowed the wind," he said of the Nazi Luftwaffe raids, "and now they are going to reap the whirlwind." It was a proclamation that proved to be as literal as it was biblical.

The Royal Air Force used Britain's early air strikes on smaller cities, along with forensic analyses of Germany's raids on England's own cities at the war's outset, as laboratories and case studies for their engineers, mathematicians, and architects to develop a more devastating style of urban bombing. Rather than trying to destroy a city with concussive blasts and shrapnel from a relatively small number of large bombs, including four-thousand-pound "blockbusters," the British researchers concluded it would be more effective to pack RAF bombers with loads of smaller explosives, some as little as four pounds. Those baton-shaped magnesium-fueled incendiaries weren't designed to blow things up. They were made to burn them down.

The fire sticks did their damage by igniting small blazes that set aflame the everyday stuff that a family might stash in an attic. Portraits. Love letters. Furniture. Baby clothes. Targeting civilians on this level might seem cruel and futile in a war being waged by millions of soldiers on three continents, but the British came to see that even a family's most intimate and mundane possessions as something militarily significant—fuel.

After a first wave of big bombs blew out an entire neighborhood's doors, roofs, and windows, subsequent waves of airplanes unleashed their loads of incendiary bombs upon the same area. The flames from these little firebombs, fanned by the drafts roaring through all the freshly ventilated homes and businesses, raged so ferociously they could quickly burn into a structure's timbers. This allowed the fires to intensify and spread down a block, the other end of which might already be on fire in a similar fashion. Harris believed if enough little fires were started on enough blocks fast enough—faster than fire crews on the ground could douse them—all the little fires could merge into a super blaze and a whole city might be reduced to ash.

Harris also liked to use a special class of thirty-pound torpedo-shaped firebombs because they could be better aimed than the smaller fire sticks that flitted from the sky with all the precision of falling oak leaves. And the flame these larger bombs made was a unique sort—once exploded, the bombs' contents splattered glowing globules that not only burned at steel-bending temperatures but also stuck like glue to anything they hit. People included. This, Harris concluded, had a "marked effect on morale of the enemy." The bombs were packed with phosphorus.

Hamburg had been harassed but largely unharmed by small-scale English air raids beginning in 1940, but by 1943 even Nazi bosses knew it was only a matter of time before the ever-swelling fleet of Allied bombers struck en masse at Hamburg's oil refineries, shipyards, and U-boat installations—and the neighborhoods that supplied their workers.

The Nazis prepared for the attack by creating a fire-fighting brigade of thousands and building more than one thousand fortified bunkers for Hamburg's 1.5 million residents.

On the first night of the week-long 1943 Hamburg attack, code-named "Operation Gomorrah," coffee dealer Nossack scrambled for safety with his wife behind their cottage's cellar door, but he eventually ventured outside and was stunned to see what looked like "glowing drops of metal" falling from the sky ten miles north in Hamburg. By the time the bombs stopped dropping fifty minutes later, Nossack described the sky to the north as red and aglow as if it were a spectacular sunset. It was 1:30 a.m.

None of the raids were as devastating as the one launched on the third night of the attack, during which English bombers hit a handful of Hamburg's cramped working-class neighborhoods with some two thousand tons of explosives, more than half of them incendiary bombs. The thousands of fires lit on that unusually hot, dry night merged in a matter of minutes into something war planners had never seen—a two-mile-wide whirlwind firestorm that burned hot as a furnace. The winds that were sucked into the cyclone to feed the oxygen-starved flames were powerful enough to topple trees three feet in diameter, ferocious enough to tear children from their mothers' arms.

British pilots that night reported seeing nothing below but

a swarm of roaring orange flame lashing from a vast bed of red coals, all of it fueling a tight column of smoke and superheated gases that mushroomed miles into the sky. On the ground, wine bottles melted. Fork tines glowed. Giant embers whipped through the city like tracer bullets on winds so loud one survivor described them as the sound of "an old organ in a church when someone is playing all the notes at once."

Civilians were hit by globs of phosphorus falling from the sky that caused their heads to burst into flames "like torches." Some jumped into canals to snuff the chemical fires but they inevitably had to come up for air, at which point the phosphorus flames flared back to life, like wicked versions of trick birthday candles.

The death toll of Operation Gomorrah was put at about thirty-eight thousand. But a precise body count was impossible, given that in many cases there were literally no bodies to count; in some instances physicians resorted to weighing piles of ash and estimating from there.


CENTRAL HAMBURG TODAY is a glassy metropolis peppered with stone and brick facades that survived the firebombing. On the streets there is little physical evidence of the horror that forced nearly one million residents to flee, but reminders do surface periodically. Literally. Some fire bombs missed their mark, and their globs of phosphorus landed in the Elbe River and its canals, where they cooled and congealed and persist on the riverbed today as harmless as a pebble, provided they stay submerged. But if one of these nubs is removed from the water and warmed to about eighty-five degrees, it will flare to life with all the ferocity it had when it hit the water in July 1943. Phosphorus pebbles also show up northeast of Hamburg on the Baltic Coast—Simanski's neighborhood—where a V-1 and V-2 rocket factory on the island of Usedom was similarly fire bombed in summer 1943, just two weeks after Hamburg.

As for formal memorials to Operation Gomorrah, a statue of a melted human form on its knees in a prayerful pose can be found today on a bustling Hamburg street. It marks the site where 370 civilians died of carbon monoxide poisoning when the fires raging above their bomb shelter consumed all the oxygen. There is also a cross-shaped patch of grass in the Ohlsdorf Cemetery near the Hamburg airport that holds the charred remains of the firestorm victims. The gravesite is marked with a statue called "Crossing the River Styx" that depicts a mother on a boat comforting her child as they float the mythical current toward the underworld, along with a few other passengers, including a naked man sitting slouched in the stern, head hanging with hands clasped behind his neck.

And just north of the Elbe River in downtown there is a statue of a barefoot man in a similarly distraught pose, but with his face buried in his hands. It sits on the site of the old St. Nicholas church, a neo-Gothic masterpiece built in the 1800s, whose 483-foot-high spire was listed as the tallest building on Earth for a couple of years in the second half of the nineteenth century, and it was still tall enough in 1943 that the night-raiding English pilots used it as a bullseye to hit the neighborhoods below. The body of the church burned to the ground in the firebombing, though its underground crypt has been restored and functions today as a museum to the carnage.

Remarkably, the St. Nicholas spire survived the attack and still scrapes the sky today. You can take a glass elevator up through its blackened core to an observation deck where a plaque attempts to explain that the phosphorus-fueled firestorm that all but destroyed the city below wasn't necessarily the fault of the Allies. The Nazi bombings of Warsaw, Rotterdam, Coventry, and London, it notes, triggered the vicious Allied response. "The many dead, wounded and bombed-out citizens of Hamburg," it concludes, "were victims of the Nazis' aggressive policies, their attempt to make Germany a world power and the barbarisation of the war that they had started."

But this plaque replaced an earlier one sometime after 2012 that was even more direct in assigning blame. That one read: "The fuse for the firestorm was lit in Germany." There evidently was some debate about whether that historical assessment was appropriately phrased. But there can be no debate, from a scientific standpoint, that the fuse for the phosphorus firebombing was indeed lit in Germany. It was, in fact, lit less than a mile from the St. Nicholas spire. Hamburg, you see, is phosphorus's birthplace.

IT WAS ONLY 8 P.M., but a full moon was already hanging high in a slate-gray Hamburg sky back in 1669 when the black magic happened. A portly man with wrinkled hands and more hair on his neck than on his head dropped to one knee in his laboratory, looked to the heavens and gestured for his two young assistants to stand back as a terrifying shaft of blue vapor shot from a glass globe balanced atop a three-legged stool.

This famous rendering of the discovery of elemental phosphorus, captured on canvas by English painter Joseph Wright a century after it actually happened, is certainly an embellishment in some respects. The artist made the real-life wizard—an alchemist named Hennig Brandt—appear much older than his actual age at the time of the discovery. He also set the scene in a cavernous hall with grand Gothic arches, columns, and massive windows instead of where it likely happened—in Brandt's home laboratory in what is today a leafy residential neighborhood near St. Michael's Church in central Hamburg, just a ten-minute walk from the St. Nicholas spire.

But the focus of the artwork—the otherworldly substance Brandt trapped in the bottle—was real. And what would have been in that glassware in the hours after it cooled and the vapors inside dissipated was a residue that cast a dazzling green-blue light. Heat did not generate this glimmer; the waxy nuggets, maybe the size of chocolate chips, got no warmer than room temperature. Yet they could stay aglow for days on end. Brandt had created something nobody at the time had ever seen. He affectionately called it "mein Feuer"—my fire.

Hennig Brandt's life up to that point had been unremarkable. He was described by one contemporary as "a man little known, of low birth, with a bizarre and mysterious nature in all he did." Born in 1630, Brandt was a veteran of the Thirty Years' War who held no great rank and left no lasting mark on the battlefield. After the war he reportedly ran an unsuccessful glass-making business before pursuing a career as a self-proclaimed physician; he signed his letters "Hennig Brandt, *Doctor of Medicinae and Philosophiae*," despite apparently having no formal education.

Brandt got rich by marrying into wealth and proceeded from there to drift deeply into the dark arts of alchemy, the ancient pursuit of gold that forged mysticism with laboratory noodling. An essential difference between alchemists and chemists, who in the eighteenth century would inherit much of the alchemists' lab equipment, experimental wizardry, and data, is that chemists are trained to seek knowledge for knowledge's sake, painstakingly acquiring it by observing, hypothesizing, and experimenting. Their methodical approach not only reveals mysteries of the material world but can also, of course, yield tremendous practical results for humanity—pulling nitrogen fertilizer from thin air, harvesting penicillin from mold, etc.—as well as wealth for the chemists who make the discoveries.

Alchemists, on the other hand, went straight for the gold. They believed that what differentiated base metals like tin and lead from precious gold was that those lesser substances had yet to evolve into a golden state, an evolution alchemists thought happened in the natural world. Alchemists believed they could give that natural metamorphosis a boost with potions and decoctions that could be distilled, precipitated, and sublimated from common materials. Turning lead into gold in this manner sounds ridiculous today, but consider how it is still commonly— and incorrectly—held that a humble chunk of coal can become a high-grade diamond if only enough pressure is applied. The alchemists' sought-after tool for similarly transmuting metals was called the philosopher's stone, the chimerical substance they believed could not only conjure gold but also cure the terminally ill and reverse the aging process.

Once the philosopher's stone was isolated, the alchemists believed, the next step was to mix fragments of it into a pot of base metal and heat it until the entire molten amalgam turned to gold pure enough to be sold by the ingot.

One ancient estimate of the power of the philosopher's stone was that, if properly enriched, a single ounce of it could convert more than seventeen thousand pounds of lead into pure gold. Some thought the magical substance could be derived from mercury or antimony or sulfur. Others sought to divine it from blood, hair, and even eggs. Brandt was a urine man.

He came to believe traces of the Stone could be found inside the human body, and he settled on liquid excreta as the likely channel to tap it. He might have been wrong that pee could turn anything more than a snowbank gold, but he was right in his hunch that our waste contains some of the most precious stuff on Earth—life-giving, life-sustaining, and life-destroying phosphorus. His discovery happened after he boiled vats of urine (presumably collected from friends and family) until all that was left was a black sludge. Then he cooked that stuff in an oven until it released a luminescent vapor, some of which condensed into the mysterious pebbles that glowed in the dark for days on end.

Brandt didn't initially share his discovery because he saw it as just one step on the path to the ultimate alchemical goal of conjuring gold. But after several years of tinkering with his discovery and no success, Brandt began selling samples to fellow alchemists who were eager to show it off to the courts of Europe, mostly as a curiosity. Yet other alchemists, having learned that Brandt had derived his glowing substance from urine, eventually cracked the recipe and began making small batches of it on their own.

The precise method for making what soon came to be known as phosphorus (Greek for Bringer of Light) remained a closely held secret for decades. Even those who did know exactly how to concoct it often failed. And those who did succeed soon learned it was not worth the risk, because the pebbles had a knack for exploding into flames hot enough to destroy laboratory equipment, and flesh.

"I don't make it anymore," said one of the first people to replicate Brandt's secret process. "A lot of mischief can come from it."

I WANTED TO replicate Brandt's wizardry by making my own elemental phosphorus, and it was not hard to find some graduate students willing to at least help me attempt the experiment. I already own a tripod outdoor propane burner, a huge metal pot, an industrial-sized thermometer, and a couple of pairs of oversized safety glasses—my turkey-frying gear. And as father of four school-aged kids and as friend to a good number of beer drinkers, I also have access to a steady stream of urine. But once I contacted an actual chemistry professor I quickly realized that, whatever you might think about the scientific legitimacy of the alchemists' pursuit of the philosopher's stone, there is no question that those phosphorus pioneers were serious laboratorians doing serious experiments in seriously dangerous work environments.

I fully realized the challenge I was up against only when I turned to an eighteenth-century recipe that provides step-by-step instructions for coaxing the element from human urine. An overly simplified summation of the incredibly detailed process: It starts with fermenting about twenty gallons of "pure" urine for several days and then cooking it until it becomes "clotted, hard, black, and nearly like chimney soot." Then you put that crust, about three pounds worth, into an iron pot and heat it until the black metal glows red and the crust stops smoking and begins to smell sweet. Then you mix in water, sand, and charcoal, and cook that sludge in a ceramic vessel at white-hot temperatures for roughly twenty-four hours, a process that, toward the end, required adding charcoal to the furnace every minute or so. What is left after some more hocus pocus are the waxen clumps of elemental phosphorus.

The folly of attempting this was made clear to me by Professor Lawrence Principe of Johns Hopkins University. He holds PhDs in both chemistry and history and has personally re-created some of the early alchemists' laboratory experiments. I asked him if he had any advice about how I might follow in Brandt's phosphorus footsteps. His email reply was cordial but stern:

> Oh my gosh, it's awful!! The problem is that one needs
> to reach red-heat to reduce the phosphates in urine to
> phosphorus, and modern glassware is simply not adequate
> for the task. Brand and others used stoneware retorts which
> aren't produced anymore. Secondly, even if one solved
> that problem, one still has the issue of condensing white
> phosphorus vapor into a solid without the entire apparatus

exploding in a ball of white fire. Yes, folks in the eighteenth century and a few in the seventeenth managed this, but the process rarely worked adequately and often enough resulted in serious or fatal injury. Only a handful of people ever managed to make it work right, and usually only those who had watched someone "in the know" do it for them first. I certainly like to redo old experiments, but this is one I think I'd pass on! (yes, I did try it once long ago, without results.)

IN THE DECADES following Brandt's discovery, phosphorus remained little more than a novelty used to dazzle kings and their courts with its bewitching ability to cast an icy glow in the dark. Even though phosphorus couldn't turn a fleck of anything into gold, it wasn't long before scientists learned how to turn it into money by selling it as medicine. It was pitched as an elixir that could trigger erections in the impotent, heal the bacteria-ravaged lungs of tuberculosis victims, tamp epileptic seizures, soothe toothaches, and lift the spirits of the depressed. Science would eventually show that elemental phosphorus can do none of these things.

But about a century after Brandt's discovery, scientists finally began to realize that the most remarkable thing about phosphorus isn't the ravenous fire it makes in a lab. It's what happens when croplands are lacking it: nothing.

# CHAPTER 2

## The Circle of Life, Broken

I N THE EARLY 1600s, pioneering chemist Jan Baptist van Helmont conducted a marvelously simple experiment into the nature of plant growth. Many at the time assumed that vegetation was the product of the soil in which it grew, that plants literally converted that loamy material into roots, stalks, and seeds. To determine whether this was true, van Helmont deposited precisely 200 pounds of oven-dried soil into a clay pot the size of a garbage can. He sprinkled in water and planted a five-pound willow sapling, and then he watched it grow. He watered the tree with either rainwater or distilled water—the purest source he could find. He also covered the pot as best he could to keep "dust" from blowing into it.

Five years later he pulled from the pot a tree that weighed 169 pounds and three ounces. Then he dried the soil in the pot to see how much of it had been consumed by the now-maturing tree. And what did he find? "The same 200 pounds, less about two ounces," van Helmont wrote in a paper published after his

1648 death. "Thus," van Helmont concluded, "164 pounds of wood, bark and roots had arisen from water alone."

Van Helmont didn't understand the process of photosynthesis that allows a plant to strip carbon from the air to add to its mass, and he evidently didn't pay much attention to the couple of ounces of soil he apparently lost along the way. But it turned out something in those missing fifty-seven grams was just as important to the tree's growth as the carbon and water the tree had consumed.

The terms *soil* and *dirt* often are used interchangeably, but dirt is a lifeless amalgam of sand, silt, and clay, dead as the moonscape. Soil contains some of that same matter, and so much more. It is a functioning ecosystem unto itself, packed with fungi and bacteria and crawling with worms and insects. It is also rich with the fertilizing elements that allow for this subterranean universe to thrive, and for greenery—everything from a blade of grass to a sky-scraping redwood—to sprout from its darkness.

Soil is not indestructible; over time it can be sapped of its life-essential fertilizing properties to the point that it literally becomes dead as dirt. This was rare before human civilizations because the fertilizing elements that a plant took from a patch of soil were returned to that soil when that plant died and decayed. Sometimes a plant took a detour into the mouth of a grazing animal, but it was eventually returned to the soil in the form of manure.

For millions of years, nomadic humans and their predecessors lived inside this virtuous circle of life. They took from the soil by consuming vegetation (or animals that ate vegetation), and they gave to the soil through their excrement, waste disposal, and ultimately, their decay. All this started to change with the advent of agriculture and the emergence of cities as humans began to grow their food in one place and eat it in another. When this had gone on long enough, the soils that sustained a community became so nutrient-degraded that famine could strike.

Humans eventually began to save their soils—and themselves—by learning to churn decaying matter back into the land they cultivated. It is not hard to see how primitive humans figured out how to repair the circle of life in this manner. People everywhere must have noticed that wherever an animal—humans included—defecated or died, the vegetation around it thrived, especially as cultures began herding livestock some ten thousand years ago. Homer even makes mention of Odysseus's beaten-down dog, Argo, lying atop a mound of mule dung that had been piled high for the farmhands to spread on the fields.

But as Europe's population began to explode in the late 1700s at the onset of the Industrial Revolution, there simply wasn't enough animal manure to keep its overworked soils productive. England was under particular stress. Its population doubled in the first half of the nineteenth century to fifteen million and was on its way to doubling again by 1900. There was no way the number of the island nation's tillable acres could keep pace with that kind of population growth, so squeezing more from existing fields was England's only option.

This forced British agriculturists to search for sources of fertilizer beyond feces. Waste shavings from the factories that made knife handles and buttons from animal bones were a particularly popular form of fertilizer in the early 1800s, so popular that it wasn't long before England started running out of animal bones.

This propelled the English into some dark places.

THE BATTLE OF WATERLOO lasted about ten hours, during which nearly fifty thousand soldiers were killed or wounded—a casualty rate of more than one per second.

Yet aside from a 140-foot high conical mound capped with

a giant iron lion to commemorate the spot where a prince was wounded (in the *shoulder*), at Waterloo today there remains scant trace of the 1815 battle that raged upon what are now waving fields of wheat and rail-straight rows of lettuce, and on the day I visited in late 2019, mountains of freshly harvested sugar beets.

A muddy dump truck collected the mounds of golden beets that, grimly, appeared about the size, color, and shape of what you would expect of a long-buried human skull. During a break in the loading I approached the dump truck operator with a question. He did not speak English (and I do not speak Flemish, French, or German), so I pulled out my phone and typed into Google Translate to ask him in French my question: "Do you ever find bones?"

He squinted to read my screen, and then his face went slack. "No," he muttered, shaking his head as his arm shot out the window to get the phone out of his hand and back into mine as quickly as he could. "No!"

It wasn't a bad question. While you can find a handful of skulls on display at the battlefield's visitor center, hardly any other human remains have been recovered since researchers started probing the site in the decades after the Duke of Wellington's troops and their allies put a bloody end to Napoleon's twelve-year roll across Europe.

"In two centuries," British historian and author Gareth Glover told me, "only a box worth of bones about one square meter in size is all that has come out of that ground."

So, what became of the remains of the thousands of men and boys who fell on the fields at Waterloo on that soggy June day in 1815?

There was a natural progression to battlefield looting during the Napoleonic Wars. It started even as the last musket and cannon balls buzzed overhead, with scavengers grabbing weapons and coins and anything else that could be pulled from a dead or wounded soldier's pockets.

Then off came the uniforms, along with badges, belts, and boots. Sometimes a soldier's head was sheared and his hair sold on the wigmaker market, but the body-stripping didn't stop there. Tooth decay was so rampant in early nineteenth-century Europe that dentists began making dentures from cadaver teeth. Battlefields like Waterloo were particularly fertile because the "donors" were typically so young their tooth enamel had yet to be pocked by sugar-rot or stained by tobacco. "Oh, Sir, only let there be a battle and there'll be no want of teeth," one scavenger replied when asked how he expected to find enough incisors, bicuspids, canines, and molars to feed London's hungry denture market. "I'll draw them as fast as the men are knocked down!"

But it wasn't until years after the last soldier hit the ground at Waterloo that the battlefield scavengers began their truly grim reaping. Newspaper reports began trickling in just a few years after the battle that some of the fertilizer used to grow England's wheat, cabbages, and livestock-sustaining turnips was not coming out of the backside of farm animals, nor was it coming from cow bones.

One English newspaper reported in September 1819 that several ships had arrived in the port of Grimsby with their cargo holds loaded with bones, which was not unusual. But what was out of the ordinary was the bones were mixed with chunks of coffin. "Those skilled in anatomy," the article stated, "have no hesitation in pronouncing many of the bones to have belonged to human beings."

In 1822 an author who identified himself only as a "living soldier" wrote a piece that appeared in London's *Morning Post* asserting that more than a million bushels of human bones, many of them fallen soldiers, were being imported to England annually. So many human remains arrived from the continent so regularly that a special "bone-grinding" mill had recently opened in eastern England to handle the imports. "It is now ascertained, beyond a doubt, by actual experiment, upon an extensive scale, that a dead soldier is a most valuable article of commerce," the soldier wrote, "and for aught that I know to the contrary, the good farmers of Yorkshire are in great measure indebted to the bones of their children for their daily bread."

By the late 1820s, at least three bone-crushing mills were operating in England, and farmers were spreading ten to twenty bushels of bones on an acre, a practice that agriculture experts of the time noted led to miraculous crop yields. Most bones were pulverized into powder, which farmers learned was the best way to get a quick fertilizing fix. Some bones were cracked down to chips, which provided a less intense but longer-lasting bump in crop productivity. Sometimes whole bones were scattered across a crop, sort of like a crude precursor to Miracle-Gro plant food spikes.

Nobody at this point knew precisely why or how bones worked to grow things like turnips and wheat, but everybody agreed that bones were helping to sustain England's nineteenth-century population explosion. "Thousands of acres are now under tillage and producing rich crops, which, but for the aid of bone manure would have continued to be mere rabbit warrens, or at most, supplying a miserable support to a few half-starved sheep," the Leicester *Chronicle* reported in 1839.

By the 1860s, however, it was becoming difficult for the British to find—to mine—enough of the dead to sustain the living. Fortunately, a fresh source of fertilizer had been found on the other side of the globe by one of the world's best-known explorers.


BARON FRIEDRICH WILHELM Karl Heinrich Alexander von Humboldt was born in Berlin to a prominent Prussian family on September 14, 1769, three months after the Duke of Wellington was born in Dublin to an aristocratic Anglo-Irish family and one month after Letizia Bonaparte gave birth to Napoleon, according to legend, on a ragged carpet in the drawing-room of her Corsican home.

Humboldt's father was an officer in the Prussian military. The notion that his son would one day follow his footsteps onto the battlefield seemed logical to those who knew the family. That included Frederick the Great, who, while visiting the Humboldt family one day, came upon the young Alexander as he studied with a tutor in the shade of a linden tree on the sweeping lawn of the family estate.

"Name?" asked the king. "Alexander von Humboldt, Sire," replied the eight-year-old.

"Alexander," the king responded. "That is a beautiful name. I seem to recall an earth-conqueror by that name. Do you wish to be a conqueror?"

"Yes, Sire," said the boy who would become one of the most celebrated scientists of the nineteenth century, "but with my head."

Humboldt grew up to make his name as an explorer and naturalist. He spent much of his life trying to synthesize the emerging fields of botany, zoology, oceanography, geology, climatology, meteorology, and mineralogy, something many people today would call ecology.

Thanks largely to the research compiled during his five-year excursion to the Americas at the dawn of the nineteenth century, Humboldt would eventually be able to describe the "netlike intricate fabric" of intertwined life that he saw wrapping the world with such clarity that one contemporary naturalist proclaimed Humboldt "the greatest scientific traveler who ever lived." His name was Charles Darwin.

Wellington has a gray piece of meat wrapped in pastry named after him. The 5' 5" Napoleon has his own psychological complex. And Humboldt? Today, in the living world alone, you can pick a Humboldt mushroom, get pricked by a Humboldt cactus, tend to a Humboldt orchid, and paint a Humboldt lily. A big-eared bat. A monkey. A dolphin. A penguin. A hog-nosed skunk. All Humboldts.

And then there is the giant Humboldt squid. It swims in what is arguably Humboldt's most important discovery, at least for the development of modern agriculture—the Humboldt Current. The massive flow of nutrient-rich and fish-filled water coursing up the west coast of South America drew Humboldt and his fellow expedition members in 1802 to a cluster of desert-dry islands off Peru, where the aridity confounded him. Winters were chronically foggy and it almost always felt like it was about to rain. But the drops never fell.

Humboldt sailed out to one island off Pisco, Peru, that he found devoid of vegetation but teeming with fish-feasting birds like terns, gulls, pelicans, and cormorants. Ever the measurer, Humboldt caught a cormorant and recorded how much it defecated in a day. Five ounces. On one of these tiny Peruvian islands alone there were an estimated five million nesting sea birds, and through their digestive tracts passed some two million pounds of fish per day. In most places on Earth, regular rains would have washed the resulting "waste" into the sea. Not in Peru, where the Andes Mountains suck away moisture before it can precipitate along the coast as rain. So the bird poop on the islands accreted over millennia, creating mountains of chalk-like dung, some more than one hundred feet high.

These "guano islands" had long been recognized by pre-Columbian South Americans as a critical source of agricultural fertilizer. The Incas, in fact, so valued the lives of the guano-making birds that, according to some accounts, the penalty for anyone caught disturbing them was death.

In the sixteenth century, the conquistadors heedlessly destroyed the Inca Empire, along with its agricultural economy that had evolved an intricate and highly productive system of canals, reservoirs, crop terraces, and guano distribution networks. But when Humboldt arrived more than two centuries later, he saw enough guano still in use by the indigenous farmers along the Pacific coast to suspect it might similarly benefit European farmers.

Despite his crews' protests about the stink, Humboldt brought home a batch of the dried bird poop to see if it could work similar miracles on the other side of the Atlantic.

Word of Humboldt's South American discoveries had spread far and wide across Europe during his 1799–1804 expedition, so much so that by the time he returned home he was a big enough celebrity to earn an audience with Napoleon in the gardens of Tuileries Palace. Europe's most famous warrior did not appear to be impressed by the famous explorer. "I understand you collect plants, monsieur?" Napoleon inquired. Humboldt replied yes. Napoleon shrugged and, before he walked off, sniffed: "So does my wife."

Small agricultural tests of Humboldt's guano yielded promising results, but one of the first farm field-scale experiments—if not *the* first—using guano as a fertilizer outside of South America was actually conducted in 1809 on a small British island in the South Atlantic. The island's governor, familiar with the guano experiments in Europe, was eager to see if locally sourced bird poop could similarly boost potato and beet crop productivity on his barren scrap of volcanic rock. Some sample plots were fertilized with guano, others with horse droppings, and yet others with pig manure. The guano-fertilized crops won big, and soon the island had twice as much land in agricultural production.

This was good news for everyone on the wind-swept island of St. Helena, including the diminutive middle-aged man who would disembark from a British Royal Ship under the cover of darkness in October 1815. He made his way to a hut-of-a-house owned by a family named Balcombe that was once occupied by the Duke of Wellington himself.

Wellington had only been visiting the island several years earlier on his way back from India. This new occupant, Napoleon, would spend the rest of his days—all 2,027 of them—as a British prisoner on St. Helena, much to the delight of the duke who vanquished the Little General on the battlefield at Waterloo only a year earlier.

"You may tell Bony that I find his apartments at the Elysee-Bourbon very convenient," Wellington wrote in 1816 to a friend on St. Helena tasked with guarding Napoleon in his exile. "And I hope he likes mine at Mr. Balcombes."

IT TOOK NEARLY three decades for the European fertilizer revolution sparked on St. Helena to spread east across the Atlantic. One reason for the lag was the nearly eight months it took for a ship to make the roundtrip from England to the west coast of South America to get the stuff. Another was that nineteenth-century Peruvians were not in a big hurry to sell off such a prized natural resource.

But throughout the 1820s and 1830s, small batches of Peruvian guano continued to occasionally arrive in England, and they continued to prove their potency on experimental crops. Finally, with English farmers on the path to exhausting animal and battlefield bone supplies, and famine worries growing for England's rapidly urbanizing population, the Peruvian government struck a deal with European businessmen in 1840 to begin regular guano shipments across the Atlantic.

The next year, some six million pounds of desiccated South American seabird feces arrived in England. The following year, more than forty million pounds were imported. Only five years after the trade started, there were hundreds of vessels shuttling almost six hundred million pounds of the stuff annually from the west coast of South America to the United Kingdom.

Much of the Peruvian guano was rich not only in phosphorus (P) but also in nitrogen (N) and potassium (K)—what we recognize today as the three key fertilizing elements. Some of the South American guano deposits, in fact, contained almost similar ratios of N-P-K that you can find in today's store-bought chemical fertilizer.

This meant guano wasn't merely a substitute for phosphorus-rich bones: it was an upgrade. During the second year of the guano trade, the Liverpool *Mercury* concluded that its impact on the nation's "over-worked worn-out soils" was nothing short of "magical." It was calculated at the time that one pound of imported guano was the equivalent of importing eight pounds of wheat.

As one publication of the era explained: "The bird is a beautifully arranged chemical laboratory, fitted up to perform a single operation . . . to take the fish as food, burn out the carbon by means of its respiratory functions, and deposit the remainder in the shape of an incomparable fertilizer."

Each bird's digestive tract was also something of a mini poison factory. An English physician reported in 1845 that he tended to a farmer who had gone to town to pick up his load of guano and, in the farmer's rush to get back home, he carelessly filled his bags with the caustic powder. "He held one corner of each of the bags in his mouth," the physician reported. "The guano was very dry, and he felt some of the dust of it going down his throat." The farmer died not long after, having vomited cups of blood.

The doctor who treated him advised guano users to be careful not to inhale puffs of the chalky substance, but the pick-swinging guano miners in Peru didn't have a choice.

Perhaps not surprisingly, Peruvian merchants had trouble finding locals willing to do the grueling, dangerous work of harvesting guano on the scale demanded by the British. Convicts were initially employed, but there were not enough of them to fill the guano cargo holds of all the waiting ships. Slave owners, meanwhile, were reluctant to risk their "property" by having them do such a dangerous job.

Eventually, mine operators turned to Chinese laborers, known pejoratively at the time as "coolies," young men so eager to escape their war-ravaged homeland that they entered into servitude in exchange for a passage to the Americas. The luckier ones arrived in the United States and South and Central America, and got jobs as cooks, bakers, gardeners, and gold miners. The unluckier ones wound up doing hard labor on the railroads or as plantation workers. The unluckiest landed on the guano islands.

Records show that of the 7,884 Chinese laborers who sailed from China to Peru between 1860 and 1863, some 2,400 perished on ocean crossing, a fatality rate greater than 30 percent. But the voyage was only the beginning of the misery. Estimates of the number of laborers shipped to Peru during the peak of the trade that lasted from 1849 to 1874 are as high as one hundred thousand. Not all of them mined guano, but those who did worked under constant threat of floggings, and many did not survive the job.

Some died from the toxic dust. Some from exhaustion. Some were killed while trying to escape. Many took their own lives; one news report of the time detailed an episode in which some fifty miners joined hands and jumped to their deaths from one of the mountains of guano.

The tales of horror that were unfolding on the Peruvian islands appeared in papers like the *New York Times* and the *Manchester Times*, but they did nothing to slow the mining of South American guano deposits that were so vast they were believed to be almost inexhaustible. In the mid-1800s, even as farmers in the United States and across Europe also began to rely on

South American guano, some predicted that the reserves on Peru's islands alone would last deep into the twenty-first century. Others said that when deposits on other South American islands were considered, the continent's guano supply was essentially "limitless."

Instead, the Peruvian deposits played out not in a matter of centuries but in years. The country exported nearly twenty-eight billion pounds (thirteen million tons) of guano between 1840 and 1880, and the supplies were largely exhausted by 1890.

Those who were there at the height of the Peruvian dung rush were stunned that they lived to see its end.

"When I first saw [the islands] 20 years ago, they were bold, brown heads, tall, and erect, standing out of the sea like living things, reflecting the light of heaven, or forming soft, tender shadows of the tropical sun on a blue sea," wrote one mining engineer in the late 1800s. "Now these same islands looked like creatures whose heads had been cut off, or like vast sarcophagi, like anything in short that reminds one of death and the grave."

EVEN AS THE guano trade exploded in the mid-1800s, many English farmers continued to rely on cheaper and more accessible caches of bones. The problem was that bones seemed to work marvelously as a crop booster in some areas of England but not in others. This set the chemists of the day to work trying to figure out just what it was about bones that gave them their fertilizing properties.

One was John Lawes, a country kid rich enough to squander his education at England's elite Eton prep school. "I learnt just enough," he once boasted, "to [avoid] punishment, but no more." After a similarly unsuccessful stint at Oxford, Lawes returned home and tried to make a go of it as a farmer on his family estate north of London. Lawes didn't apparently have a gift for Greek or Latin, or for literature or art, or for philosophy or math. But he did somewhere along the way acquire a tinkerer's interest in chemistry. He began to explore how chemicals might enhance crop yields after a neighbor suggested he conduct experiments to figure out why their own soils failed to respond well to bone fertilizer when bones worked such wonders on farms in other regions of England.

Toiling in a lab he converted from a barn, Lawes came to understand the problem was that the soils in his region lacked a natural acidity needed to unleash the fertilizing powers of bones, as well as certain types of rocks. He demonstrated this, to fantastic effect, by blending these materials with sulfuric acid and feeding the admixture to cabbages grown in pots during experiments in the late 1830s.

By 1840 he had cranked up his experiments to include whole fields. The bigger the scale, Lawes reasoned, the better. He wasn't content to have his work shown merely in the scientific literature; he was eager for his neighbors to see it in action.

"He wanted the experiments done large," Paul Poulton, a retired soils scientist told me one gray November afternoon in 2019 as he bounced his muddy black Ford Focus down rutted roads cut along the same fields Lawes had worked almost two hundred years earlier, "so he could demonstrate the effectiveness of fertilizers to other farmers, and so they could see it work on their scale."

Lawes patented his acid-bone concoction in 1842 and branded the product "superphosphate." He made so much money from this pioneering chemical fertilizer that he eventually donated his estate so it could be converted into a giant agriculture laboratory known today as Rothamsted Research, home of the oldest ongoing agricultural experiments in the world—and a striking monument to the power of chemicals to keep intensively farmed fields fertile over decades, even centuries. The real-world ramifications of the fertilizer research at the time were stunning; one study revealed that between 1840 and 1880, the average yield for a crop of cereal in England nearly doubled.

But even as Lawes was starting to work wizardry in his fields with chemically enhanced bones, scientific rivals toiling in their own laboratories were headed in similar directions. That included Germany's Justus von Liebig, who came to the same conclusion about enhancing bones' fertilizing power by mixing them with acid. But Liebig did most his work in a lab, not a farm field.

Considered by many to be the founder of organic chemistry, Liebig burned plants and found they released carbon, oxygen, hydrogen, and nitrogen—all elements that can be found in abundance in air and water. In the ash he also isolated, among other elements, phosphorus and potassium. From this research, Liebig in 1840 popularized what became known as the theory of mineral plant nutrition, arguing that you didn't need to derive fertilizers from something that was once living. You could start with the raw materials themselves—lifeless elements. The dawn of the chemical fertilizer revolution was cracking.

IF YOU WANT to make five ham and cheese sandwiches, you are going to need ten pieces of bread, five slices of ham and five slices of cheese. If you only have eight slices of bread, you can only make four ham and cheese sandwiches. If you only have two pieces of ham, you can only make two ham and cheese sandwiches. And if you don't have any bread, you can't make any ham and cheese sandwiches, no matter how many slices of ham and cheese you have.

This is, of course, common sense, but the law of the "limiting factor" was revolutionary when Liebig and others began to apply it to fertilizing croplands. The idea is that crop growth is not capped by the sum total of *all* the various types of soil nutrients that a plant needs, the three most essential ones being, as we now know: phosphorus, nitrogen, and potassium. Growth is instead limited by the *least available* of those nutrients.

One way that professors demonstrate what is known today as the law of the minimum is to ask their students to picture a wooden barrel filling with water. A barrel is commonly built with thirty individual bowed planks, called staves, lashed together by steel loops at the barrel's bottom, middle, and top. Now, if one of those staves is seven inches too short to reach the barrel top and another stave is two inches too short, you will never fill the barrel more than seven inches from the top because, of course, the barrel's shortest stave is its limiting factor. Now, if you fix the stave that is seven inches too short so it reaches the barrel top, the barrel can then be filled further, but only up to two inches from the top. That's because the second short stave is now the limiting factor in filling the barrel.

The law of the minimum removed an element of voodoo from growing a crop by establishing that farmers didn't necessarily *need* bones, or cow manure or guano, or hair, or blood or marl or anything else that they found, through trial and error, sustained crop production. Instead, they needed the essential soil nutrients—P, N, and K—that those natural fertilizers contained. And with advances in chemistry the idea was that agriculturists could, through soil sampling, figure out what nutrient was most lacking in a field and then remedy it in a manner that bumped up crop production. Liebig went so far as to proclaim that a time was coming when fertilizer prescriptions could be written for individual fields "exactly as at present medicines are given for fever and goiter."

He was correct. But writing a prescription is one thing. Filling it was another matter altogether.

By the 1850s, even with the guano trade taking off, English

farmers were still importing almost any clavicle, femur, tibia, and patella they could get their hands on. The scavenging, which included plundering Egyptian ruins for their human remains and mummified cats, left Liebig aghast.

"Great Britain deprives all countries of the conditions of their fertility," Liebig fumed. "It has raked up the battle-fields of Leipzig, and Waterloo, and the Crimea; it has consumed the bones of many generations accumulated in the catacombs of Sicily. . . . Like a vampire it hangs upon the breast of Europe, and even the world, sucking its life-blood without any real necessity or permanent gain for itself. It is impossible to imagine that such a sinful disturbance of the divine order of things should be allowed to go on forever with impunity and the time will probably arrive for England, earlier even than for the rest of Europe, when, with all its wealth in gold, iron, and coal, it will not be able to repurchase the thousandth part of those conditions of life so frivolously wasted for centuries past."

But new discoveries were about to make bone fertilizing all but obsolete. Several years after John Lawes filed his 1842 chemical fertilizer patent, he reworded it so his recipe didn't even specifically mention treating bones with sulfuric acid. The patent was modified to basically include any materials that might contain useful amounts of phosphorus. And, it would turn out, that included far more than bones and bird poop.

Our natural world would never again be so natural.

# CHAPTER 3

# *Bones to Stones*

A s the "inexhaustible" Peruvian guano deposits played out near the end of the nineteenth century, farmers were forced to scour the globe for not only new sources of phosphorus but also the two other fertilizing nutrients.

Potassium was the relatively simple fix; accessible deposits of it remain abundant even today in the form of salts left behind by long-dried-up seas. Nitrogen was another story. Deposits of a mineable form of nitrate (a compound of nitrogen and oxygen) were harvested during the nineteenth century in some desert regions of South America, but agriculturally useful geologic caches of the element were globally scarce.

It isn't that there is a lack of nitrogen in nature—the air we breathe is more than 78 percent nitrogen. The problem is that it's in a form unavailable to most plant life, much the way all the O in $H_2O$ does a drowning boy no good.

But there is a family of plants that converts atmospheric nitrogen into the form needed by crops—legumes. Peas, beans, peanuts, lentils, clovers, and all their cousins can (in simple terms) pull nitrogen from the atmosphere and bind it with hydrogen atoms. So farmers could recharge a nitrogen-depleted field by regularly planting crops of legumes and then plowing them back into the soil to supply nutrient-starved, non-nitrogen-binding crops like wheat, rice, and corn.

By the end of the nineteenth century, agricultural scientists understood the value in planting legumes to keep fields fed with nitrogen. But they also fretted that legumes alone couldn't sustain enough cereal crops to keep up with the ballooning human population. Something was needed to replace the dwindling South American guano and nitrate deposits.

Then along came a miracle, in the form of an accused war criminal—one of the most villainous—and esteemed—scientists Germany ever produced.

FRITZ WAS NOT the prototypical World War I front-line soldier. He was bald and bespectacled, thick in the hips, slow in the trenches. And his attention on the battlefield always seemed to be fixed more on the breezes overhead than on the bombs and bullets whizzing past. Fritz held a degree from a prestigious university but entered the military as a middle-aged man with the enlisted rank of sergeant major. Yet it was Fritz who called the shots on the fields of Flanders on April 22, 1915. More specifically, it was Fritz who ordered German troops to hold their traditional fire and instead attack in an unexpectedly dastardly manner.

"It was a beautiful day," one of the German soldiers who followed Fritz's orders would later recall. "The sun was shining. Where there was grass, it was blazing green. We should have been going to a picnic, not doing what we were about to do."

What they did was open the valves on some five thousand pressurized gas cylinders that let loose a lazy haze along four miles of Belgian battlefront that then drifted toward the Allies on the winds Fritz so carefully monitored. The Allies on the other side of no-man's-land thought the gray-green cloud rolling toward them was a smokescreen, a prelude to a savage enemy charge, but it turned out the cloud *was* the charge. It was chlorine.

"What we saw was total death. Nothing was alive. All of the animals had come out of their holes to die. Dead rabbits, moles, rats and mice were everywhere," the German soldier recalled. "When we got to the French lines, the trenches were empty but in a half mile the bodies of French soldiers were everywhere. It was unbelievable. Then we saw that there was some English. You could see where men had clawed at their faces, and throats, trying to get their breath. Some had shot themselves."

One estimate of the toll of the first large-scale chlorine gas attack in history put the number of dead at more than one thousand and another seven thousand injured. By the end of the war, the chemical attacks unleashed by both the Germans and the Allies—but initiated by Fritz, leader of Germany's chemical warfare program—would claim 1.3 million casualties on both sides, including up to one hundred thousand deaths.

In the months following the war's end in 1918, Fritz became infamous around the world as one of Germany's most heinous war criminals. But the next year he was known as something altogether different: winner of the Nobel Prize.

Both reputations are deserved because Fritz's legacy could not be more ambiguous. An army of German scientists and technicians might have worked their chalkboards and laboratory gear to develop bigger bombs, deadlier guns, stouter tanks, and faster planes. What separated Fritz was his eagerness to slip out of a white lab coat and into his baggy army uniform (replete with the Germans' signature spikey Pickelhaube helmet) to personally orchestrate battlefield slaughter.

Yet the same hands that were stained with the blood of thousands of Allied soldiers also worked miracles to save untold millions more civilians from starvation and paved the way for Earth's population to balloon from 1.6 billion in 1900 to more than 7 billion today. What, specifically, did Fritz do to deserve that Nobel Prize?

Fritz Haber figured out how to make bread from thin air.

He essentially made the world's nitrogen fertilizer scarcity problem disappear on July 2, 1909, when he demonstrated an invention that could do the work of a thousand fields of legumes by converting normally inaccessible atmospheric nitrogen ($N_2$) into fertilizing ammonia ($NH_3$). He used heat, immense pressure, and a metal catalyst to cleave hydrogen atoms from methane and bind them with atmospheric nitrogen to create plant food.

His crucial discovery was made even more important in 1913 after fellow German chemist Carl Bosch figured out how to scale up the process so it could be done on an industrial level, fortunate timing for the German war machine that also depended on ammonia for ammunitions.

What's now known as the Haber-Bosch process is just as essential to humanity today as it was in the early 1900s, even more so. As the journal *Nature Geoscience* put it in a 2008 article: "The lives of around half of humanity are made possible by Haber-Bosch nitrogen."

Yet Haber's discovery did not render moot Liebig's law of the minimum because it did nothing to address the supply of phosphorus. That bottleneck was cracked in a completely different manner in the middle of the nineteenth century, and it was done with help from a young woman who could wield a hammer like no one else.

◊ ◊ ◊

THE MAIN HALL of London's Natural History Museum is not only a monument to the wonders of the natural world but also to Imperial English masculinity.

On the landing of the Romanesque chamber's grand staircase is a towering marble statue of Charles Darwin. To Darwin's left hangs a portrait of Alfred Russel Wallace, the other father of evolution. To Darwin's right stands a bronze statue of a rifle-cradling, slouch hat–wearing Captain Fredrick C. Selous. The famed British naturalist (and lion killer) who died in Africa in 1917 at the Battle of Behobeho is sometimes called "blood brother" to the most swashbuckling of American presidents, Theodore Roosevelt. Nearby in the cavernous hall is a bronze relief of a couple of legendary (male) ornithologists.

Aside from a plaque from Queen Elizabeth II commemorating the museum's one-hundredth birthday back in 1981, the only prominent female character I could find in the whole scene on the day I visited loomed from the rafters of the seven-story-high chamber. She is the nine-thousand-pound skeleton of a great blue whale. She was harpooned off the east coast of Ireland in 1891. Her name is Hope.

I had to pick my way down a hallway spoking off the main chamber to find mention of a nineteenth-century female naturalist where, across the way from the museum's "T-Rex Grill," there is a portrait of a stern-looking woman in a frumpy green dress. She is clasping a hammer. The painting's title: Mary Anning—*the fossil woman.*

Anning, born in 1799, became famous long after her death as the alleged inspiration for the children's tongue-twisting poem *She Sells Sea Shells by the Seashore.* But Anning did more than peddle shells. She was a fossil excavator *extraordinaire,* a premier paleontologist before the word even existed. She was, the eminent science historian and evolutionary biologist Stephen Jay

Gould once said, "probably the most important unsung (or inadequately sung) collecting force in the history of paleontology."

Anning wasn't even ten years old when she and her older brother Joe began scouring the coast near their English seaside town of Lyme Regis for remnants of ocean creatures frozen in rock and time, but their dangerous excavation work along the base of constantly collapsing cliffs was anything but play. The kids were hustling to put food on the table. Their father was a struggling cabinet maker with a troublemaking streak—the year after Anning was born he led a mob in a "bread riot" to protest food shortages tied to England's grain shortages that were, at least in part, tied to deteriorating soil conditions.

The family turned to selling fossils out of their cabinet shop as means to make ends meet. It was an enterprise that dismayed the children's mother, but Mary Anning's unmatched skill at unearthing ancient life not only helped keep her family fed—it would, in a twisting way, also help put Britain on a tack away from the chronic fertilizer shortages of the eighteenth and early nineteenth centuries, a path that would, along with Haber's invention, help make way for billions of more humans to inhabit the planet.

It all started with the giant head of a toothy beast that the hammer-swinging kids chipped from a cliff. That head today is on display at London's Natural History Museum alongside Anning's portrait. It has an alligator-like snout the size of a PGA player's golf bag, teeth thicker than cigars, and eyes the diameter of dinner plates. As mesmerizing as it is menacing, the specimen nevertheless is squeezed onto a recessed shelf along the hallway in a manner that leaves it unnoticed by the legions of school children who stream past on their way to the cafeteria, oblivious to the real-life monster over their shoulders that is big enough to swallow them whole.

Joe Anning apparently had enough after extracting the rocky

skull. But Anning, trained by her father to patiently tease away the cliffs' flaking shale, spent the next year leading a group of Lyme Regis workmen extracting more of the animal's fossilized remains.

The beast she unearthed (her father died before it was fully excavated, after taking a tumble off a nearby bluff and then contracting tuberculosis) is called an ichthyosaur, which is derived from Ancient Greek and translates into fish lizard. Looking like a cross between dolphin and alligator, the reptiles could, by some accounts, grow to more than eighty feet and make their way through the water at motorboat speed.

Like whales, ichthyosaurs were air-breathing descendants of land creatures that, for whatever reason, took a deep breath and headed back to sea. Also like modern-day killer whales, ichthyosaurs' backs were far darker than their nearly luminescent white bellies—exactly the color scheme you would expect for a beast that attacked from below.

Anning would go on to find other sea creature remains that she sold to museums and to tourists, and at some point she unearthed a specimen so intact that she found what she thought was evidence of fossilized dung in its fossilized digestive tract. And, like a Russian nesting doll, these nuggets appeared to hold their own trove; when Anning cracked them with her hammer she found they were packed with fossilized bones and scales.

Anning began to make a name for herself in England's burgeoning community of naturalists, despite the fact she had no formal training in the sciences. That doesn't mean she was uneducated. She taught herself by transcribing copies of the research papers shared with her by scientists who came to watch her work, all the way down to their exquisitely detailed illustrations. Wrote one prominent Londoner in 1824 who met Anning in Lyme Regis when she was still in her mid-twenties:

"The extraordinary thing in this young woman is that she has made herself so thoroughly acquainted with the science that the moment she finds any bones she knows to what tribe they belong. She fixes the bones on a frame with cement and then makes drawings and has them engraved. . . . It is certainly a wonderful instance of divine favour—that this poor, ignorant girl should be so blessed, for by reading and application she has arrived to that degree of knowledge as to be in the habit of writing and talking with professors and other clever men on the subject, and they all acknowledge that she understands more of the science than anyone else in this kingdom."

So when Anning concluded she had found evidence of extinct sea creatures' diets, the learned men of the time listened.

THE IDEA OF fossilized feces went mainstream in the late 1820s after pioneering Oxford University geologist William Buckland, who had worked on excavations with Anning and praised her "skill and industry," presented to the Geological Society of London his argument (bolstered by Mary's discoveries) that the pinecone-shaped stones regularly found amid the remains of ichthyosaurs and other ancient sea life were indeed fossilized stools. He named the rocky droppings "coprolites," derived from the Greek words for feces: *kopros* and stone: *lithos*, and he reported that they weren't confined to the stomachs of petrified sea creatures; in some coastal areas of England they could be found on their own, tons of them, littering the landscape "like potatoes scattered on the ground."

Buckland described the earthy clumps as resembling, well, run-of-the-mill poop.

"They, for the most part, vary from two to four inches in length, and from one to two inches in diameter. Some few are much larger, and bear a due proportion to the gigantic calibre of the largest ichthyosauri . . . some are flat and amorphous, as if the substance had been voided in a semifluid state," Buckland wrote. "Their usual colour is ash-grey, sometimes interspersed with black, and sometimes wholly black. Their substance is of a compact earthy texture, resembling indurated clay."

This was all more than a scatological curiosity. That ichthyosaurs feasted on other sea life—and even their own young—upended the Christian belief of the time that God's creatures lived in harmony before the Fall. It even inspired an 1830 painting that brought to life many of the fossilized species Mary had unearthed in a fashion that graphically illustrated the kill-or-be-killed reality for prehistoric life on Earth. The painting depicted all manner of savage creatures, including ichthyosaurs, long-necked plesiosaurs, and belemnites (our Baltic Sea beachcomber Gerd Simanski's favorite find) chasing and preying upon each other.

The animals rendered in watercolor are either attacking with their mouths agape or desperately flitting away. Above the waterline, a fat turtle plunges from the shore to attack a squid-like creature, and a nearby alligator, standing along the shoreline with its mouth open like a barking dog, is attacked from the water by a long-necked plesiosaur. Palm trees bend in the wind as winged pterosaurs engage in a paleo dogfight in the sky above.

The Garden of Eden, the fossilized poop proved, was no picnic.

"Coprolites form records of warfare, waged by successive generations of inhabitants of our planet on one another," Buckland observed.

The painting was so popular the artist had reproductions made, and he donated proceeds of their sales to Mary to allow her to keep digging, which she did until breast cancer killed her at age forty-seven.

◊　◊　◊

IN THE EARLY 1840s, just as John Lawes was beginning to test his chemically enhanced fertilizers on crops at Rothamsted, Buckland scoured the British coast examining coprolites with two renowned chemists, Lyon Playfair and Justus von Liebig. Buckland's companions didn't only look at the sausage-shaped clumps as mere paleontological wonders but also as a potential source of fertilizer in a world that had finite caches of dried bird poop and was fast running out of bones.

"The interesting question arose as to whether these excretions of extinct animals contained the mineral ingredients of so much value in animal manure," chemist Playfair recalled some years later. "We took specimens, in order to confirm by chemical analysis, the views of the geologist."

Liebig himself conducted the analysis and was stunned it revealed the coastline was littered with coprolites and other rocky materials that were, in aggregate, loaded with phosphorus. Liebig proclaimed these rocks as, potentially, even more important to nineteenth-century England than the chunks of coal that fueled the steam engines powering the Industrial Revolution. Just as burnable coal was known at the time to be the remnant of ancient plant life, Liebig believed these fossils could be similarly harnessed for an equally important fuel—fertilizer to make food.

"What a curious and interesting subject for contemplation!" Liebig exclaimed after his coprolite analysis yielded phosphorus. "In the remains of an extinct *animal* world, England is to find the means of increasing her wealth in agricultural produce, as she has already found the great support of her manufacturing industry in fossil fuel—the preserved matter of primeval forests—the remains of a *vegetable* world."

Not everyone was as enthusiastic.

"I well recollect the storm of ridicule raised by these expressions of (Liebig)," Playfair wrote several years after the discovery, "and yet truth has triumphed over scepticism, and thousands of tons of similar animal remains are now used in promoting the fertility of our fields. The geological observer, in his search after evidence of ancient life, aided by the chemist, excavated the extinct remains which produced new life to future generations."

CERTAIN TYPES OF rocks had been mined for fertilizer on a relatively small scale prior to Liebig publishing the theory of mineral plant nutrition in 1840, but perhaps more important than the phosphorus contained in the coprolites (most of which actually turned out to be phosphorus-rich sedimentary rocks) is that the discovery of coprolites seems to have helped focus agriculturists in their hunt for greater deposits of rock-based phosphorus.

They ultimately realized, through chemical analysis, that they could find significant concentrations of phosphorus in certain sedimentary rock formations created by all manner of dead sea life raining upon the ocean floor year after year, eon after eon. Under the right conditions, the phosphorus in all that petrified detritus got concentrated as ocean currents stripped away other elements of the rock. Over the course of millions of years, geologic spasms then heaved the phosphorus-laden rock onto land, where it could be mined.

Many of the early deposits of what came to be known as "phosphate nodules" were found scattered across England, where mining of it peaked in the 1870s. The deposits were quickly depleted, and harvests had plummeted by the early 1890s, right about the time the Peruvian guano reserves were also running out.

All this was happening just as Earth's population was on its way to doubling to two billion in little more than a century. Fortunately for all those additional mouths, similarly mineable phosphorus-rich sedimentary rock deposits were found in the southeastern United States, first in South Carolina in the 1860s and then, on a far grander scale, in central Florida in the early 1880s. By the mid-1890s, scores of Florida companies and thousands of miners were harvesting more than one million tons of phosphorus rock annually.

The Florida "Bone Valley" deposits consisted largely of sedimentary rocks in the shape and size of pebbles you find underneath playground swing sets. And, as on the coast of England, they were commonly found among the fossilized remains of all manner of long-gone creatures—sabretooth tigers, humungous sharks, monstrous manatees, and super-sized bears. But the scientific significance of this bizarre menagerie was overshadowed by all the prospectors flooding into Florida with a Wild West mentality that had prospectors willing to kill over the gravel used to construct Bone Valley roads. As the Jacksonville Florida *Times Union* reported on February 13, 1890:

"Pete Downing pulled out his gun and said he owned more of the phosphate that had been thrown in the street than any two men and he meant to protect his share. . . . And they kept pulling out their guns and knives until there were 30 or 40 ready for business, each one swearing he owned most of the rock and meant to have it or the red gore would run deep."

Florida phosphorus mines were flooding the world with chemical fertilizer by the turn of twentieth century. But humans' hunger for the element only grew more acute as the often-violent hunt for it spread across the globe.

Soon the casualties weren't just individuals, but entire cultures.

BAKER ISLAND IS a scrubby scrap of rock not a whole lot bigger than a golf course that sits almost atop the equator in the middle of the Pacific Ocean. An American firm mined its easily accessible guano deposits from 1858 to 1879, at which point the company deemed the island satisfactorily exploited and sold the mining rights to an English outfit called the Pacific Phosphate Company.

With most of the guano gone, the new owners began mining the island's sedimentary phosphorus rock deposits, most of which could be excavated with the swing of a pick. But some of the phosphorus rock on Baker Island was so dense it had to be dynamited into chunks dense as a doorstop—literally.

Albert Ellis was working in the Pacific Phosphate Company's Sydney, Australia, laboratory one day late in 1899 when he noticed that a rock being used to prop open a door had a striking resemblance to some of those dense Baker Island phosphorus rocks. He mentioned this to a coworker and was told that the rock did not come from Baker Island, and that company geologists had already concluded that it was just an old heavy rock.

"This seemed conclusive enough, but somehow when at work in the laboratory that piece of rock would catch my eye and its resemblance to Baker Island would recur," Ellis recalled years later. "It was probably three months later that the thought came to test it. A bit was knocked off, ground up, and the usual tests for (phosphorus) were made."

The analysis revealed the doorstop contained some of the most concentrated phosphorus rock ever discovered, richer in the nutrient than even some Peruvian guano. The problem was the rock came from a Pacific Island that had already been claimed by Germany, but a colleague told Ellis of another island 160 miles to its east that was unclaimed by any Western power and something of a twin to the German island, in terms of its geological history. Its name on nautical charts at the time was as nondescript as you'll find on a map: Ocean Island.

Ellis hastily made plans to cross the 2,600 miles of nothingness between Sydney and that 2.3-square-mile patch of rock and coconut trees just south of the equator. "If Ocean Island is what I think it is," Ellis wrote in his diary, "there is a fortune on it, if not several."

One of Ellis's colleagues offered some harsh words for what was in store for him upon his landing on the tiny island that had acquired an outsized reputation among seafarers. "The Ocean Islanders are hard cases," he warned. "You take your rifle and your revolver with you, and as soon as you get on the beach show the natives you can use them."

MANY HAVE MARVELED about the derring-do of the French voyageurs who paddled birch bark canoes across North America's sea-sized Great Lakes in the seventeenth century. Lake Superior alone is about the size of Maine, and the east-to-west sailing distance stretches some 350 miles. But open-water crossings are not how the voyageurs typically did their business. The explorers, guided by Native Americans, got to the far sides of the "Sweet Water Seas" by paddling along their shorelines, dipping their cups in the surf along the way to slake their thirst by day, and gathering at night around campfires to feast on fish and fill those cups with whiskey.

Contrast that with the ancient Pacific migrants who set out from their home islands to hunt for new patches of land in the middle of an endless ocean. It required crossing hundreds or maybe even thousands of miles of open water—water they obviously could not drink—under a roasting sun and over treacherous swells in vessels fashioned from materials like wooden planks lashed together by twine stripped from coconut shells.

The pilots of these sail- and paddle-powered vessels had their own navigational tools—the sun and moon, the stars, winds, waves, currents, clouds, and birds. Nevertheless, too many of the attempted migrations certainly ended not with a happy landing but with empty water jugs.

Yet there were enough successful migrations that distinct cultures were thriving when the populated islands were stumbled upon by European traders and whalers in the eighteenth and nineteenth centuries. Ocean Island, which had been inhabited for at least two thousand years by the time it was "discovered" in the early 1800s, is one such place.

One of the earliest recorded arrivals of whites on Ocean Island happened in the early 1850s, when the crew of an Australian vessel anchored offshore.

The ship's crew found some of the inhabitants wore necklaces made of human teeth, but the islanders quickly made it clear to their guests they weren't hunting for fresh molars. Pleasantries were exchanged over the next few days, along with waterfowl (from the islanders) and some tobacco and a hatchet (from one of the ship's crew). The big boat soon sailed on, bringing with it a few islanders eager to see the world beyond their beach. They were lucky to get out when they did.

Ocean Island receives roughly seventy inches of precipitation annually (that makes it wetter than any city in the continental United States). Yet the landmass is so small and sunbaked that it has no perennial streams or ponds. Its only water source is what falls from the sky. So when the rain doesn't come, trouble soon does.

For centuries, islanders survived these dry spells by slithering into muddy caves to fill coconut shells with the turbid water that collected some one hundred feet below the island surface. During one multiyear drought that hit in the early 1870s, even these underground reservoirs began to dry up. Island leaders limited households to one coconut shell of water per day, but even that was too much of a draw on their reserves.

As the drought stretched into its third year, islanders were forced to suck the juices out of seaweed, to little avail.

"People's gums rotted in their mouths; their teeth fell out and their bodies were covered in ulcers," one island survivor recalled. "They fell in the pathways and died there; and where they died their bodies remained, for who was strong enough to carry corpses home for burial rights?"

By the time the rains returned in the mid-1870s, as many as three-quarters of the island's approximately two thousand residents had perished.

Another disaster struck barely a decade later, one from which none of the islanders, ultimately, would escape.

ON MAY 3, 1900, Ellis's ship arrived at Ocean Island, today known as Banaba Island. Despite his colleague's warning that the island's inhabitants, the Banabans, posed a grave danger to visitors, Ellis found them friendly and eager to trade for their shark-tooth swords, fruits, and fish.

As the swapping carried on aboard his ship, Ellis slipped off with some field-testing gear to probe the island interior for the hoped-for rock phosphorus. He figured before he left that if he found a cache of ten thousand tons of phosphorus rock it would be a "bonanza" for his company, which was on the brink of bankruptcy. By the time he finished his hasty survey that first day, he was convinced island reserves could top six million tons. "At last we had 'struck oil,'" Ellis recalled, "and never was a 'gusher' more welcome or more opportune."

Before sunset on that first day, Ellis negotiated for rights to mine the island rock with a small group of Banabans whom he believed to be the island's political leaders and therefore, he later claimed, possessed the power to sign over island mining rights. The sides "negotiated" through an interpreter who had only a rudimentary understanding of English. This might explain why the islanders accepted a handwritten contract that gave Ellis' firm the rights to mine the island's rock for 999 years. In exchange, the Banabans would, collectively, receive 50 pounds per year, or just about $8,000 in today's currency.

That first year, some 1,500 tons of phosphorus rock was taken off the island. The next year, exports ballooned to 13,350 tons, and they grew exponentially from there as the island harbor was deepened to accommodate bigger boats to carry away all the crushed rock mined by the army of laborers recruited from as far away as Japan, China, and Hawaii.

The 1900 contract was eventually revised, but the islanders still didn't get close to a fair price for their precious rocks; between 1900 and 1913, the Pacific Phosphate Company made a profit of 1.7 million pounds. The Banabans were paid less than 10,000 pounds.

As some Banabans began to resist ceding their personal land to the ravenous mining company, there was talk of kicking the Banabans off the island altogether. "It is inconceivable," the *Sydney Morning Herald* wrote in 1912, "that less than 500 Ocean Island-born natives can be allowed to prevent the mining and export of a (product) of such immense value to all the rest of mankind."

The company and the islanders eventually reached yet another deal that called for Pacific Phosphate to pay a greater rate to strip-mine additional island acres and greater royalties per ton for rock harvested. The money did not go straight to the islanders but to a fund to be managed for them. The mining company further agreed to stop forcing Banabans to pay exorbitant rates at the company store that had been gouging them on prices for canned corned beef, fish, sugar, tea, rice, biscuits, and, particularly cruel for a people who had recently suffered so immensely from drought, drinking water.

In 1920, the governments of Australia, New Zealand, and England turned the private mining enterprise into a public one run by the three governments. That same year, a visiting journalist reported how the island had gone, in just two decades, from an isolated land largely untarnished by the modern world to one of its most industrially plundered. He described how the crash of waves, the chirps of birds, and the rustle of palm branches had been drowned out by the din of a "miniature city" that roared around the clock and calendar.

"Day and night there are the mighty crash and whir of machinery," the *Victoria Daily Times* reported, "the shrill shrieks of locomotives, the deafening rattle of trucks rushing along the railways loaded with the precious phosphate for the crushers, dryers and bins, where it is treated and made ready for the tramp steamers which carry it to every nation actively engaged in agriculture."

Japan invaded Banaba Island during World War II with the idea that it had lots of use for the island's phosphorus but little use for the islanders, whom they starved, beheaded, shot, and electrocuted. Those who weren't murdered were dispatched from the island to work in labor camps. At the end of the war, the Allies collected the some seven hundred surviving Banabans from various Pacific islands and moved them to a remote island of Fiji that had been purchased as a resettlement colony—with the islanders' own mining royalties.

The Allies transported the Banabans to that island some 1,600 miles to the south of their homeland and left them with barely a two-month food supply. The flapping canvas tents the Banabans were given offered scant protection when storms predictably hit, and dozens perished in that first year.

Meanwhile, with the war over and the Banabans removed, the English, New Zealanders, and Australians returned to Banaba Island, and the pace of the strip mining picked up until reserves were exhausted in the late 1970s. Some of the last shipments of phosphorus came from the island golf course, one of the few patches of the island the phosphorus miners had shown any interest in protecting during their nearly eight-decade ransacking.

By 1980, all that was left was a mid-ocean ghost town of rusting warehouses, crumbling asbestos-roofed homes, deserted vehicles, and a decrepit steel conveyer belt that stretched across an island reef and out into deep water. A trickle of Banabans nevertheless began to recolonize their island in the decades after, and today the island's population hovers around three hundred. There is no airstrip on the island and no major industry to speak of; access to the outside world is limited to a supply boat that arrives every few months and typically only stays for a night or two.

BANABA'S ROCK PHOSPHORUS, along with similar deposits on a handful of other tiny, far-flung islands across the Pacific and Indian Oceans, was shipped all over the world during the twentieth century. But much of the rock found its way onto the remarkably nutrient-starved fields of Australia and New Zealand.

These Pacific phosphorus rock deposits are a big reason why those two island nations went from nineteenth-century backwater British colonies to twentieth-century economic and cultural powerhouses. The phosphorus-driven greening-up of the two countries led not only to richer, meat-based diets for their citizens but also to greater national wealth due to food exports to North America, Europe, and the Middle East.

"The conversion of Australia and New Zealand into mirror images of the British Isles and Anglo North America did not happen naturally," writes fertilizer historian Gregory Cushman. "It required the systemic destruction of several tropical islands to remake the soils and biota of these southern lands."

New Zealanders today rely on airplanes and helicopters to help spread a staggering two million tons of fertilizer across their countryside annually, forests included.

So when the Pacific islands' phosphorus reserves began to play out in the second half of the twentieth century, New Zealanders were desperate to secure a fresh phosphorus fix. And they found one, in what might be the only place on the planet more distressed than Banaba Island.

# CHAPTER 4

## *War of the Sands*

SATURDAY, JUNE 16, 2018, was US astronaut Drew Feustel's day off. As commander of the International Space Station, he didn't have a lot of places to go, other than around and around the world every ninety minutes. So the veteran of three space station tours and avid amateur photographer floated over to the Russians' module of the orbiting observatory to press his Nikon D5 against one of their portholes.

The Russian windows were smaller than those of the majestic "cupola" observatory on the US side of the space station, but the cosmonaut glass was of higher optical quality. This mattered to Feustel because his space hobby is snapping pictures of auto racetracks all over the globe on race days, and on this particular day was the running of the legendary 24 Hours of Le Mans. As the space station hurtled toward France at more than seventeen thousand miles per hour, something down in Africa jumped out at Feustel, who holds a PhD in geological sciences.

"I like to find tectonic features, folds and thrust belts, places where you have converging plate boundaries and where you see folded features because of glaciation," he told me of his time spent gazing down upon his home planet. But the Etch A Sketch–like picture emerging from the sand some 240 miles below didn't look like one of those geological formations. It vaguely resembled some sort of massive insect scratching its way to the surface of the Sahara Desert.

The image, which was like nothing Feustel had ever seen, became crisper as he zoomed in with the Nikon. "It was pretty easy to see it wasn't natural," he said, "mainly because of the lineations and the squared-off features." Feustel snapped a picture, and what his 1600 millimeter lens captured, human rights activists will tell you, is one of the world's largest active crime scenes. A separate image captured by a satellite arcing some four hundred miles above the West African coast just one week earlier filled out the picture.

The satellite spied a 650-foot-long freighter filling its cargo holds with what appeared to be a mountain's worth of sand tumbling from a conveyer belt atop a pier that stretched nearly two miles out into the Atlantic's turquoise waters. From the perspective of space, the scene made little sense; it was as if a ship was bothering to steal away with a stash of sand from a desert so vast it dwarfs the continental United States.

But pull back on that Google image of the ship until land comes into view, and you can see a rail-straight line across the desert running some hundred kilometers inland to the heart of the bizarre landscape formation photographed by Feustel.

It turns out the picture Feustel unwittingly captured is of a massive phosphorus rock mine built back in the days when the region was a Spanish colony. And the stripe on the ground running between it and the freighter is the world's longest conveyer belt, built a half century ago to get the chalky phosphorus rock from the mine out to the North Atlantic, and from there to farm fields around the globe.

55

It is also an emerging battle line.

Spain relinquished its colonial claims to the region, known as Western Sahara, in the mid-1970s, and the territory—along with the mine—was immediately taken over by neighboring Morocco.

Today, Morocco operates and profits off the mine, but who actually owns one of the most coveted sweeps of desert in the world, Saudi Arabia's fabled Ghawar oil field included, is at the heart of an increasingly violent dispute between Morocco and the region's native Sahrawis.

"I had no idea there was so much connected to that place," said Feustel.

He isn't alone. For centuries much of the world thought of Western Sahara as little more than a wasteland in which indigenous nomads competed for survival with horned vipers, rodents, and scorpions. Then, just as World War II was spreading to neighboring Morocco and Algeria, a Spanish geology student came along on a camel.


MANUEL MEDINA BEGAN surveying Western Sahara as a PhD student at the University of Madrid in the early 1940s. His expedition was more than an academic exercise. Even though Spain remained essentially neutral during World War II, its twenty-seven million residents nevertheless suffered from extreme shortages of natural resources because of the 1930s Spanish Civil War that gutted the country's economy.

Medina was among a group of geologists that Spanish dictator Francisco Franco dispatched to what was then Spain's Saharan colony to probe its sand and dirt for desperately needed natural resources like oil, coal, iron—and phosphorus. In those pre-GPS, pre-Land Rover days, camelback travelers like Medina crossed the endless dunescape like boats bobbing on ocean swells. There are so few geographical markers on the world's largest nonpolar desert that caravan leaders, like mariners, relied on sextants and the stars to navigate the no-man's-land between A (a place with enough water to survive) and B (the next place with enough water to survive).

Equipped only with crude field tools of the time—including rock hammers and magnifying glasses—Medina homed in on the flinty black rock of Western Sahara's ancient riverbeds and read the geologic history they contained like chapters in a book. One of the stories they told was that Western Sahara had once harbored a massive sea and that the sedimentary rock formations left on that ancient seabed were strikingly similar to a rich phosphorus deposit in neighboring Morocco that had been intensively mined since the 1920s.

Franco dispatched a battalion of scientists to zero in on the heart of the Sahara phosphorus deposit, which, legend has it, was finally discovered in the early 1960s beneath a single tree poking from a sandy sweep so devoid of features that nomads used it as a terrestrial navigation buoy. Once the extent and nature of the deposit was defined, geologists calculated they had struck upon a phosphorus deposit that was among the largest and richest on Earth.

By the early 1970s, just as the Banaba Island phosphorus deposits were nearly played out, Spain had invested some $400 million in developing a mine at the site that was so remote a German firm was brought in to design the world's largest conveyer belt to get the rock out of the desert and onto freighters at a specially built pier on the North African coast. The first load of phosphorus rock rumbled down the belt on its way to Japan in 1972. Within a few years the mine employed some 2,600 workers.

Spain saw the mine as a boon to both its own economy and to that of the native Sahrawis. The Sahrawis viewed it as an outright heist, and began to launch military attacks on the conveyer belt. Franco didn't have another fight in him. Spain negotiated an exit from Western Sahara in 1975 that resulted in Morocco taking control of the region around the mine, and the mine itself, even though Morocco does not have internationally recognized rights to the territory.

A half century later, the United Nations still doesn't classify Western Sahara as an independent state, nor does it formally acknowledge Moroccan claims to the land. It instead describes Western Sahara as a "Non Self-Governing Territory in the process of decolonization."

It is proving to be a long and bloody process.

WHO SHOULD CONTROL Western Sahara has been in dispute since King Hassan II of Morocco hurried Spain's 1975 departure by sending 350,000 of his subjects across the border waving crimson Moroccan flags, copies of the Koran, and posters of the king himself. The king claimed he was only patching the cultural rupture that resulted from the nineteenth-century Spanish occupation of a region his government claims has been part of Morocco "since the dawn of time."

Most of the Moroccan citizens that the king dispatched to Western Sahara turned around when they got there and headed straight for home. But thousands of soldiers who flanked them on their march south did not, and they unleashed a bloodbath on Sahrawi resistance fighters in what became one of the longest, most lopsided wars you probably have never heard of, at least not yet.

It's been an unfair fight since the beginning. Morocco at the time of the invasion had a population of twenty million. The Sahrawis numbered somewhere between fifty thousand and one hundred thousand, roughly half of whom—mostly women, children, and the hobbled—fled for makeshift tent cities in neighboring Algeria.

For Morocco and the tens of thousands of soldiers it dispatched to Western Sahara, the fight wasn't just over land it claimed as its own, nor was it specifically tied to potential profits from the phosphorus mine. It was a business move—Morocco has its own massive phosphorus deposits and mines and in the 1970s so influenced the world phosphorus market that it was able to, OPEC-like, set phosphorus rock prices on the global market. This would not be possible were it in competition with the Western Sahara mine.

"The Moroccan takeover is thus less important in terms of mere possession than in King Hassan's increased ability to keep (phosphorus) prices up through limitation of production," a Canadian newspaper reported in 1976. "He now controls about 80 percent of the world phosphate trade."

This "War of the Sands," as it was sometimes referred to by the few journalists who covered the conflict at its outset, flickered and flared for fifteen years until the UN brokered a 1991 cease-fire. (A 1960s conflict between Morocco and Algeria was also often called the war of the sands.) Today, that fragile peace is fraying as guerrilla warfare heats up along a 1,700-mile-long, ten-foot-high wall of sand and rock that splits Western Sahara in two. On the Atlantic side of the Moroccan-built barrier is what Morocco still calls its "Southern Provinces" and what the Sahrawis still call their stolen home. The mine and the fertile fishing grounds along the Atlantic coast also make it the most economically valuable portion of Western Sahara.

Patrolled by Moroccan soldiers operating out of posts fortified with artillery and armored with millions upon millions of land mines, the berm today is the longest militarily active partition on the planet.

That firepower has kept the mine operating more or less continually since the 1980s and allowed Western Sahara's wealth to be spread around the globe. Fertilizer made from Saharan phosphorus long boosted US crops of soy. It has been scattered behind tractors working India's muddy fields of pigeon peas, wheat, and millet. It has fueled crops of barley, potatoes, rice, and rye in Europe. It spurred Mexican corn stalks to grow tall as trees.

But in recent years, human rights groups supporting the Sahrawis have successfully pressured European and North American fertilizer companies to stop buying the phosphorus from Morocco. Even so, their boycott has done nothing to end the Sahrawis' exile, and rumors of war are beginning to swirl in the tent camps.

The violence during the Western Sahara mine's first half century of operation never made much news outside of Africa. But the trouble roiling the desert today is worthy of the world's attention. It is a window into the challenges facing civilization in the next half century as countries waken to the fraught prospect of Liebig's law of the minimum, playing out on a global scale.

UNLIKE MANURE, which is manufactured daily, and unlike guano, deposits of which can be replenished over the course of years or decades, the phosphorus rock deposits that sustain the world's modern agriculture system do not regenerate on a human time scale. This will, eventually, likely pose a problem for every person on the planet, either in the pocketbook or in the stomach.

Western Sahara's sedimentary phosphorus rock deposit, like others, was formed by dead organisms showering down upon an ancient and long dried-up seabed. Geologists note it takes millions upon millions of years for all that dead life to accrete on the ocean floor to form phosphorus-rich sedimentary rock, and for random deposits of that rock to then make their way back onto land via geologic uplift. This means that once humans blow through the existing phosphorus rock reserves, there is no reason to expect endless and easily accessible sources to miraculously appear. It is a reality that has long spooked some of the most powerful politicians in the world, including the president of the United States.

"I cannot overemphasize the importance of phosphorus not only to agriculture and soil conservation but also to the physical health and economic security of the people of the nation," the president proclaimed after noting the United States' primary supply of phosphorus rock in Florida was fast dwindling just as exports from the Sunshine State to Europe were increasing. "It is therefore high time for the nation to adopt a national policy for the production and conservation of phosphates for the benefit of this and coming generations."

The president's warning came just at the time of year when US crops of corn and wheat, soybeans and vegetables were hitting their peak fertilizer demand—May . . . of 1938.

Franklin Roosevelt's exhortation came at a time when the planet was home to some two billion humans, before the first sprout poked from the Earth in the Green Revolution, the late twentieth-century explosion in yields of essential crops like rice, wheat, and corn. The revolution that cranked up the entire planet's metabolism had its roots in the 1950s and was brought about by a combination of high-yield seed strains, modern irrigation systems, and heavy doses of phosphorus- and nitrogen-based fertilizer.

The Green Revolution started just as the United Nations calculated that more than half of humanity was in danger of starvation. It not only saved the lives of untold millions across the Third World but it then allowed the globe's population to nearly double between 1970 and today.

Liebig's law of the minimum dictated that uncapping the limitation on nitrogen meant phosphorus supplies had to keep pace, and thanks to mining they have—allowing fertilizer production to increase six-fold between 1950 and 2000. Yet the path we are on means that the pace of phosphorus extraction is going to have to keep increasing—in barely a generation, an additional two billion humans will wake up expecting to eat every day. And now, with countries across the globe moving toward meat-rich diets that demand planting of ever more acres of grains, some agriculture experts predict the Earth must double its crop production capacity again by the year 2050. This will not happen easily.

By 2019, phosphorus mines around the world—in the United States, Algeria, Australia, Brazil, China, Egypt, Finland, Israel, Jordan, Kazakhstan, Mexico, Peru, Russia, Saudi Arabia, Senegal, South Africa, Syria, Togo, Vietnam, Morocco, and Western Sahara (along with a handful of small operations in other countries), were collectively scraping from the Earth some 250 million tons of phosphorus rock annually.

It is a pace of extraction the Earth ultimately cannot sustain. Estimates for the day the reserves of easily minable phosphorus rock play out have ranged from a matter of decades (highly unlikely) to several centuries. But the problem is not just a matter of running out of phosphorus rock completely. Some phosphorus experts say trouble will loom when Earth hits its "peak phosphorus" moment—when demand starts to outstrip supply as the deposits diminish, ore quality degrades, and harvesting costs climb. And as the globe's phosphorus deposits go, so do we. "Life can multiply until all the phosphorus is gone and then there is an inexorable halt which nothing can prevent," famed author Asimov proclaimed the year before Morocco invaded Western Sahara.

The United States today is particularly vulnerable. Some phosphorus experts say it could face domestically mined shortages before the end of this century, and perhaps sooner. The idea of energy security for three hundred million people suddenly looks like an easy problem to solve, at least compared to food security.

"Phosphorus," Gary Albarelli, director of information programs for the Florida Industrial and Phosphate Research Institute, once told me, "is a hell of a lot more critical than oil."


STARVATION IS HARD to fathom in the United States, where food remains relatively cheap, even for many of those who meet the government definition of poor. The poorest 20 percent of the US population still only spends about 16 percent of its income on food. That leaves some room to trim other spending before stomachs start to ache. It's a different situation in places like Vietnam, Nigeria, and Indonesia, where food purchases can eat up three-quarters of a family's income.

In other places, food costs gobble up *all* of a family's budget, and more; despite the Green Revolution, an estimated three to four million children under the age of five succumb to malnutrition-related deaths annually. These children live in places where, World Bank president Robert Zoellick warned in April 2008, there already "is no margin" in the fight against starvation. Barely a week later, ten thousand Bangladeshis swarmed their streets to protest wages that weren't keeping pace with the rocketing price of rice. Police fired tear gas canisters to turn back attacks on authorities, textile factories, and public transportation systems.

Food supply riots were in fact rampant across the globe in 2008, when prices for corn, wheat, and rice nearly doubled in barely twelve months. Hunger-driven unrest roiled Egypt, Cameroon, Indonesia, and Haiti, where five people were killed in food protests that included a storming of the presidential palace. "For months, Haitians have compared their hunger pains to 'eating Clorox' because of the burning feeling in their stomachs," the Associated Press reported. "The most desperate have come to depend on a traditional hunger palliative of cookies made of dirt, vegetable oil and salt."

The 2008 spike in food prices—and the cost of phosphorus fertilizer—was blamed on ballooning demand for more meat in rapidly developing countries like China and India. Crop-crippling storms, soaring oil prices, and a surge in the use of grain-based ethanol to supplement US and other wealthy nations' gasoline supplies were also factors.

Stomachs full, most Westerners paid little attention to these events, many of which unfolded on the other side of their world. Yet they happened in the same world that will soon be spinning through space with nine billion souls aboard.

If you think of the globe as a life raft drifting through space— and you should—a nutrition deficiency anywhere on the planet is a reason for concern for everyone on the planet.

A life raft with even just one starving survivor, after all, is a life raft that is safe for no one aboard.


BILLIONAIRE BOSTON-BASED INVESTOR Jeremy Grantham has made a lot of money predicting bad times ahead. He warned of the bloated Japanese equities market in 1989, the tech bubble in 2000, and the housing crash of 2008. But none of those financial meltdowns come close to the disruption he expects could happen in the decades ahead as the world comes to terms with its dwindling phosphorus reserves.

"What happens when these fertilizers run out is a question I can't get satisfactorily answered and, believe me, I have tried," Grantham says. "There seems to be only one conclusion: their use must be drastically reduced in the next 20–40 years or we will begin to starve."

Grantham's dire warning did not appear in an annual newsletter to investors or in a business publication like the *Financial Times*. It ran in the scientific journal *Nature*, the same pages where Einstein presented his theory of relativity, where DNA pioneers Watson and Crick unveiled the famous double helix, and where the world met Dolly, the first cloned mammal. While Grantham's warning evidently got the attention of the editors of the renowned scientific journal, it was not well received in the financial world, or by the mineral extraction industry.

Tim Worstall, a fellow at the conservative Adam Smith Institute, shot back with a patronizing—and amusing, and thoughtful—rant against Grantham's assertions.

Worstall's counterpoint, which appeared in *Forbes Magazine*, was based on technical mining terms: reserves versus resources. Reserves are mineral deposits that have been geographically defined and deemed mineable given the limits of existing technologies and economics. Resources are, on the other hand, an overall estimate of what is known to exist on all of Earth. Resources can become reserves if a mining firm or government undertakes the expense of finding and defining new deposits.

"You'll not be surprised to learn that drilling and sampling and sending odd hairy geologists over the hill with little hammers is expensive. So we only do this with the stuff we're likely to dig up in the next few decades. Thus, reserves of ore are, at any one time, good only for a few decades of use."

In other words, Worstall was arguing that the reason Earth's phosphorus reserves appear to be running so perilously low is, ironically, because there are still enough of them to last generations.

"Grantham may indeed know how to make money," Worstall said. "But I really would suggest a technical dictionary the next time he wants to talk about mine reserves and the availability of resources. For this is just an awful, schoolboy type, error of his."

Pedro Sanchez, director of Columbia University's Agriculture and Food Security Center, agrees that worries that the Earth is about to run out of phosphorus deposits are overblown. "In my long 50-year career," says Sanchez, "once every decade people say we are going to run out of phosphorus. Each time this is disproven. All the most reliable estimates show that we have enough phosphorus rock resources to last between 300 and 400 more years." Sanchez went on to explain that technologies are evolving to make phosphorous mining more efficient. He is also confident there are vast potential reserves on the ocean floor that will one day make their way onto our croplands.

But what is missing in this back-and-forth about phosphorus scarcity is the reality that the Earth doesn't have to run out of phosphorus rock for life to be disrupted across the planet.

Phosphorus deposits are not spread evenly across countries, let alone continents. Most of them, as we've seen, lie inside the borders of Morocco and Western Sahara, collectively home to roughly 70 to 80 percent of the world's reserves. It is a hyper-concentration of a globally essential resource that Grantham called "the most important quasi-monopoly in economic history."

The United States, for example, has very roughly one billion tons of phosphorus rock reserves remaining (unbridled development in Florida is a significant obstacle to any effort to try to expand its reserves) and it is mining some twenty-five million tons of them each year. This puts the richest country in the world at risk of running out of existing phosphorus reserves in thirty or forty years, after which it could become dependent on other countries to keep its population adequately fed. And while countries all over the world, including Algeria, Australia, Brazil, Egypt, Jordan, Kazakhstan, Peru, Russia, and Tunisia, are scrambling to develop their own relatively small reserves, it seems quite possible that at some point Morocco is destined to be the world's primary phosphorus dispenser. And it would not be just one country in control of this life-essential element. It would be one family, even one guy.

Shares of the company that owns Morocco's phosphorus mines, including the one it has claimed in Western Sahara, are largely held by the government, which itself is controlled by the king of Morocco, Mohammed VI, also known as M6.

If there were any doubt about who is in charge of the world's largest cache of phosphorus reserves, it evaporates when you turn to the first page of one recent annual report for the government-run phosphorus fertilizer company that happens to be one of Morocco's largest economic enterprises. It's a portrait of M6, captioned: "His Majesty Mohammed VI May God Glorify Him." Under M6's rule, you can be imprisoned for speaking ill of Islam, for speaking ill of the king, or for engaging in homosexuality.

"This share of reserves makes OPEC and Saudi Arabia look like absolute pikers, and phosphate is much more important even than oil," Grantham proclaimed in 2018, not long before news broke that two female Scandinavian hikers were beheaded by Islamic extremists while vacationing in Morocco.

"If ISIS takes over Morocco, I give my . . . personal guarantee that within a week the military of China or the US or both will have intervened," says Grantham. "We simply cannot manage for long under currently configured agriculture without Morocco's reserves, perhaps 35 to 40 years."

AMONG THE SAHRAWI refugees who sought shelter from the tanks and machine-gun fire during the 1975 Moroccan invasion of Western Sahara were the grandmother and mother of Najla Mohamedlamin, who is now in her early 30s. Najla's mother was six years old at the time her family made the dash to a tent camp safe zone on the Algerian side of the border. Najla's mother and siblings were told they would have to stick it out in the hastily erected compound for several weeks, maybe even a month or two, before tensions settled and they could return home. But then four months turned into four years, and then four years turned into four decades.

Today the family still lives in the camp, still sleeps in an olive drab canvas tent and still survives on fifty-kilogram sacks of rice delivered by aid workers from the United Nations World Food Program. The family's toilet is still a hole in the ground. Its drinking water still arrives by the jug.

There are now an estimated 125,000 Sahrawis living in this manner in a cluster of camps inside Algeria. Generations of Sahrawis have grown up—and grown old—in the tent cities. Their economy, if you can call it that, is dependent on international aid, even as the Moroccan-operated Western Sahara phosphorus mine hums around the clock less than a day's drive to the west, doing more than a quarter billion dollars in business annually, at least before human rights activists started their boycott campaign.

When Najla was eight years old, she believed the whole globe was desert, and that everyone lived in a world where a cup of spilled water drew a stiff scold. Then a humanitarian group brought her to Spain for summer camp. "Oh my god," she told me of her first time living outside her tent camp. "When I first saw a swimming pool I just thought: 'What?' Water to us is a very, very precious thing. And if you waste water you are going to get into big trouble. And then to see people standing and *playing* in all this water?"

Then there were the televisions and automobiles, the shopping malls and the food, particularly the juicy green watermelons with their almost neon pink insides. "You think to yourself," Najla said, "'All this exists?' And then, from that time on, I started to realize there is something wrong here."

When Najla returned to her tent, she was committed to getting an education, first in a small school at a neighboring camp and then during extended stays in Austria and Spain. By 2016 she had made her way to a community college in western Washington state, and in 2018 she earned an associate's degree.

Her goal after graduation was to get back home, not just to the tent camp but, she hoped one day, to the lands Morocco invaded nearly fifty years ago.

She says a big obstacle is the phosphorus mine that, along with coastal fishing rights, is why Morocco invaded Western Sahara, even while Morocco has long maintained its goal in occupying Western Sahara is to reunite related peoples. "If it was just a desert with nomads and nothing else, who wants that?" Najla said. "Our sin is that the land is rich—in phosphate."

Geopolitical screws have been turning on Morocco in recent years from human rights activists pressuring countries and companies to stop purchasing what they see as the king's ill-gotten trove of phosphorus. The campaign is working. In 2012, more than $300 million worth of Western Sahara phosphorus rock was reportedly sold around the globe, and customers have since been dropping rapidly. By 2018 one of the only buyers left in the Western world—maybe *the* only buyer—was New Zealand.

"It is my country's wealth that is turning your soils green," Najla wrote in 2018 in an open letter that appeared in the New Zealand medial outlet *Stuff.* "Every time the malnourished refugee receives donations from the UN, we think of you. And we hope you spare us a thought the other way. Know that we are poor so that Morocco can be rich, by transporting our national wealth to harbours, far, far away."

The goal of this "Blood Phosphate" campaign is to pressure Morocco to pull out of Western Sahara and turn the mine over to the Sahrawis in a move that would give them both the freedom to return to their ancestral lands and an economic starting point.

Whether Morocco's subjugation of the Sahrawis is—or isn't—resolved offers a glimpse of the power of the phosphorus atom on an increasingly cramped globe, in both constructive and destructive ways. Phosphorus will either bring nations together by flowing across political divides, whether they are only lines on a map or physical barriers as impenetrable as the Western Sahara berm and vast as the oceans. Or it will tear them apart.

Najla is not optimistic.

"I am very worried," she told me in 2018, shortly before returning to her Algerian camp, "that war will be the only solution in the end."

In 2021, armed Sahrawis resumed commando-style attacks on the berm protecting the mine.

*Part II*

# THE COST

# OF

# PHOSPHORUS

CHAPTER 5

# Dirty Soap

Twelve-year-old Charles Frosch was taking a shortcut home on a chilly morning in March 1956 when he toppled off a stone wall and plunged into the frothing Baraboo River in downtown Reedsburg, Wisconsin. A friend tried to toss him a line, but Charles, weighed down by his winter jacket, couldn't reach it before he slipped under. Rescuers arrived by the dozens and scoured the river channel, first with their eyes and then with the hooks of corpse-grabbing draglines in what turned out to be an excruciating recovery effort.

Soon dozens of searchers from area fire and police departments descended on the scene. Hundreds of gawkers lined the riverbank, hands stuffed in coat pockets, waiting for the boy's body to surface. The search for Charles wasn't just hampered by gusty winds, biting temperatures and ice floes, but also foam. Soap foam. So much soap foam that it smothered the river channel from bank to bank.

Firemen failed to blast away the froth with their high-pressure hoses. Crews piloting amphibious World War II surplus "Duck" transport vehicles tried vainly to push away the suds. Even sticks of dynamite didn't do the trick. The US Air Force finally dispatched helicopters to the river, the idea being the pilots could dip their blades, just so, to fan away the foam, but the river kept gurgling bubbles faster than they could be blown away.

The only trace of the boy anyone found in the days after he disappeared was his winter cap.

The Reedsburg police chief finally located Charles' remains nearly two months later in a tangle of debris just yards from where he slipped under the surface. News accounts from the time noted that a rosary would be said for the boy on April 22 at a Reedsburg mortuary and that a funeral would be held the next day at nearby Sacred Heart Catholic Church.

Oddly, none of the media coverage identified the source of the bubbles, nor did any of the reports indicate that a frothing river was anything out of the ordinary. This was, after all, the mid-1950s—the dawn of the washing machine detergent era, a time in which foaming rivers and lakes had become, almost overnight, only natural.

But there was nothing natural about the super-potent boxes of synthetic soap suddenly flooding supermarket aisles in the 1950s. One chemical used in the new detergents, it turned out, made bubble messes of the rivers, lakes, and oceans that were on the receiving end of all the dirty water whisked away in washing machine spin cycles.

Another critical element in the new synthetic soaps had a more devilish side effect. It didn't just pollute waterways. It burned the life out of them. It was phosphorus.


THE WORDS *soap* and *detergent* are often used interchangeably today, but the two cleaning agents are almost as far apart on the technology spectrum as a horse-drawn buggy is from a Tesla.

Soap has for thousands of years been made from animal fats mixed with lye-producing ashes to create molecules that have an exceptional ability to pull grease and grime from skin, hair, and clothing.

The remarkable power of the soap molecule comes from the fact that one end of it is hydrophilic, which gives it the ability to make water "wetter." It does this by breaking the glue-like bonds between water molecules that cause water to bead up. This allows soapy water to work its way into the minuscule cracks, crevices, and seams that are the unseen three-dimensional world of a piece of cloth, a strand of hair, or a patch of skin.

A soap molecule has another cleaning muscle on its tail end, which is lipophilic—drawn to microscopic particles of oil and grime and quick to bind with them. So soap's "wetter" water gets the cleaning started by setting afloat bits of grease and dirt. Then the filth-craving ends of the soap molecules smother those suspended flecks at the same time the water-loving ends of the soap molecules remain attached with similar vigor to surrounding water molecules. The result is that all the scrubbed-away dirt, grime, and bacteria-rich globules stay afloat in the water so they get flushed away before they can re-bond to the materials from which they were scrubbed.

Soap was hand-mixed in the home for centuries, but by the mid-1800s it was being churned out on an industrial scale. One of the biggest manufacturers was Cincinnati-based Procter & Gamble, which made money during the Civil War by converting the fatty wastes from the city's slaughterhouses into candles and soap. (So much meat was butchered in the southern Ohio city, known at the time as Porkopolis, that it was said its canals ran red with pig blood.) As much as soldiers needed bullets, boots, and blankets, after all, they needed candles to navigate the night. They also needed soap—way more than they knew. For every Civil War soldier killed on the battlefield, two more died from disease. Diarrhea and dysentery alone felled tens of thousands of soldiers in a war that was over before the medical world grasped microbes' role in causing and spreading lethal diseases, and soap's power to fight them.

The emergence of lightbulbs in the decades after the war might have ravaged P&G's candle-making business, but there was still loads of money to be made from the humble bar of soap, and no business capitalized on the public's craving for cleanliness like P&G.

In the 1800s, soap was typically sold butcher-style—in slabs carved on demand by apron-wearing merchants who wrapped them in brown paper and charged by the pound. P&G changed that in the 1870s when it began selling uniform bars that came prewrapped. More importantly, there was a brand name on those wrappers, and soon enough that name was plastered on buildings and billboards across the country—Ivory.

Most soaps at the time were of similar quality. Ivory was distinct in that air was whipped into the formula during the manufacturing process. This meant if a bar were dropped into a pond, bathtub, or wash basin full of turbid water, it would, conveniently, bob like a cork. It was a feature that hatched one of P&G's first signature slogans—"It Floats"—but it wouldn't be the last. P&G's uncanny ability to use branding to dominate the market with its high-end cleaning products would prove fantastic for America's consumers, and eventually, tragic for America's waters.

Even with Ivory's success, P&G's soap business was headed for trouble by the 1930s as motorized washing machines began to make their way into basements across the United States. The appliances could, its manufacturers claimed, eliminate a day's worth of drudgery with the push of a button. The problem was the automatic washers did not clean nearly as well as a vigorous hand-scrubbing, particularly when the machines were used in homes served by mineral-rich hard water, whose magnesium and calcium blunted the machines' effectiveness.

To capitalize on the washing machine boom, P&G sought to develop a high-power synthetic cleaning agent specifically engineered for washing machines. "This may ruin the soap business," company president William Cooper Procter warned as his researchers began tinkering with various chemical cleaning formulas. "But if anybody is going to ruin the soap business it had better be Procter and Gamble."

The company began selling its first synthetic product in the 1930s, but it still wasn't strong enough to get the job done. So company chemists focused on enhancing their concoction with a "builder" chemical that could make the detergent more potent by, essentially, adding to the formula a water softener that could neutralize the hard-water minerals. The special element in that builder? Phosphorus. Or, more specifically, sodium tripolyphosphate.

The problem was that clothes washed with this phosphorus-boosted synthetic detergent came out of machines satisfactorily clean but annoyingly stiff and crusty, so the chemist in charge of what P&G referred to as "Project X" worked to come up with a formula that had just enough builder to get clothes clean, but not so much it left them cracker-crisp. He made scant headway until he took a different tack; instead of trying to mix as little phosphorus into the detergent as possible, he went in the opposite direction and super-dosed the formula with phosphorus. Counterintuitively, clothes washed with this formula came out clean *and* velvety soft.

At the time, P&G scientists didn't know exactly why their phosphorus-rich formulation worked so well; they just knew it did. And if anyone pondered the potential harm of pushing millions of pounds of a phosphorus- and petroleum-based detergent onto American grocery store aisles and then flushing it down basement drains and into waterways across the country, it wasn't the P&G marketers. They were, naturally, focused on the job of convincing American consumers the company had once again struck upon an Ivory-like, revolutionary product.

The P&G advertising men (at the time the job of marketing was almost exclusively male) knew the ladies doing all the washing (at the time the task of cleaning clothes was largely handled by women) loved to see bubbles in their wash. The more bubbles the better, so much so that, one day not long after P&G decided to go to market with Project X, its inventor lost his cool after he was interrupted one too many times as he tried to walk members of the company advertising department through the nature of his product and what made it so much better than ordinary soap.

*But what about the bubbles?* they kept asking.

The exasperated researcher finally cracked. Oh, he told them, his stuff made bubbles alright. In fact, he said, a box of it could make "oceans of suds." Another slogan was born. The next step was to get the public hooked on this new concoction. They started by calling it *Tide.*

Instead of focusing only on traditional ads in newspapers and billboards, the P&G marketers also bought airtime, first on radio and eventually television, to tell original, fanciful stories for their target audience—chore-stressed women spending the days of their lives with all their children while pining for another world.

Media critics of the era didn't refer to these specially produced dramas sandwiched between detergent pitches as television shows or radio programs. They called them soap operas, and by the early 1950s P&G had become the biggest advertiser in the nation, spending some $45 million on marketing annually, much of it on what fast became known among the public as "soaps." The strategy worked; only five years after Tide was introduced in 1946, P&G and its competitors were selling a billion pounds of synthetic detergent annually. Clothes across America were coming out of the washing machine softer and whiter and brighter than had ever been possible with traditional soap.

But the dirty business of getting clothes clean didn't really go away. It just flowed downstream.

IT DIDN'T TAKE long for scientists to trace the bubbles that started plaguing US waterways in the 1950s back to suds in the washing machines. The problem was that the cleaning agent in Tide and its rivals was based on an oil derivative that, unlike the traditional soap molecules it replaced, was not easily digested by natural microbes swarming in open waters.

Suddenly blobs of bubbles were blizzarding off rivers in clusters so thick they caused car crashes. One bubble mass on Illinois's Rock River crested almost five stories above the riverbank. These bubbles were so durable and so widespread that by the early 1960s they weren't just making their way through sewage treatment systems and into streams, rivers, and lakes. They were bubbling back into the homes that produced them, gurgling through the pipes of public drinking water systems that drew from the same waters into which the suds were discharged. News accounts from the time told tales of tap water so thick with the synthetic froth that people were doing their dishes with water straight from the faucet.

Tide's inventor might have been sarcastic when he promised to make oceans of suds, but soon political leaders on both sides of the Atlantic were accusing detergent makers of literally making suds of the ocean.

US Congressman Henry Reuss took a trip to Europe in 1962 and could barely comprehend the size of the mess synthetic detergents had so quickly made of Denmark's North Sea. "At Elsinore, where Prince Hamlet confronted the ghost of his murdered father, on the rampart overlooking the sea, I saw what seemed to be either the ghost's ectoplasm or a gigantic iceberg come down from the north," Reuss testified before his congressional colleagues upon his return. "By all the logic of oceanography, there could not be an iceberg there, and sure enough, there was not. The iceberg was a mountain of foam floating serenely along on the water."

Reuss was more than a tourist on the trip. The Harvard Law graduate who helped put Europe back together after World War II as deputy general counsel for the Marshall Plan knew that German scientists had succeeded in formulating a new "soft" detergent that did not produce such absurdly durable bubbles, and he went to their laboratories to see for himself the production of this new "biodegradable" detergent.

Reuss soon introduced legislation requiring US detergent makers to convert to the new formula, but the detergent industry, having already invested untold millions of dollars convincing the public it could not live without its bubbles, argued it was too late to go back.

"There is nothing so difficult to handle as an irate housewife. And they have a strong impact on the industry. We might develop an effective detergent with no suds at all, but that wouldn't do," a spokesman for the National Soap and Detergent Association told a group of sanitation engineers at a 1962 University of Minnesota symposium on the bubble conundrum. "Imagine trying to convince a housewife she could wash without suds."

Yet the new formulation that was widely adopted in Germany by 1964 still produced loads of bubbles. The difference was they popped once they left the washing machine. Under pressure from an American public growing increasingly skittish about the environmental price being paid for their washday miracle, US detergent makers converted to the new formulation in 1965, and the nation's bubble troubles dissipated almost immediately.

But the havoc synthetic detergent inflicted on the environment didn't fade with the disappearance of foam on rivers and lakes. Lurking beneath all the white fluff, it turned out, was a far graver problem tied to the detergents—a continent-wide explosion of putrid green algae that was growing so thick on lakes and rivers across the country that by the mid-1960s it was choking the life out of them.

ECOLOGISTS HAVE THREE basic classifications for lakes, in terms of how much aquatic life they can support.

*Oligotrophic* lakes are filled with crystalline waters and, because they lack an abundance of nutrients, are populated with relatively small amounts of plankton and fish. Lake Tahoe and Lake Superior fall under this classification.

*Eutrophic* lakes are on the other end of the spectrum. They are typically nutrient rich, warm, murky, and teeming with fish due to a profusion of plankton at the base of the food chain.

*Mesotrophic* lakes, including Germany's Lake Constance and Bolivia's Lake Titicaca, are in between the clear state of an oligotrophic lake and a soupy eutrophic lake.

A eutrophic lake is often not long for this world. Its ever-reproducing, ever-dying plant and animal inhabitants relentlessly accrete on the lake bottom, eventually squeezing out spaces for fish and other aquatic animals, and over time, water itself. At some point a eutrophied lake becomes more swamp or bog than lake and, ultimately, it disappears altogether into the landscape.

It is a natural aging process that can take tens of thousands of years or more. At least it was natural until humans began to supercharge algae blooms in the twentieth century with their industrial chemical discharges and poor to nonexistent treatment of sewage.

No lake suffered more from these human-triggered algae

blooms than Lake Erie in the mid-1900s, when thousands of square miles of lake were proclaimed "dead" because its pollution-fueled algae inevitably died and decayed, sucking so much oxygen out of the water that almost nothing could survive.

Things got so bad that newspaper columnists at the time wrote Lake Erie eulogies. "It will always be there to look at and for ships to sail on," wrote one Pennsylvania editor in 1966, "but it may soon be dead except for the algae and sludge worms." Added an editor from Ohio: "Can you imagine the size of Lake Erie and not marvel at the fact that we Americans have joined forces with Canadians to make it a dead sea?"

Dr. Seuss sealed Erie's dead-lake reputation in 1971 in *The Lorax*, in which he wrote of a world with waters so polluted it drove fish ashore. "They'll walk on their fins and get woefully weary in search of some water that isn't so smeary," Seuss wrote. "I hear things are just as bad up in Lake Erie."

The big question among scientists, politicians, and business leaders of the time: Precisely what type of nutrient pollution was driving this deadly explosion in life? This was a critical question because, Justus von Liebig's law of the minimum dictated, if you isolated the one nutrient that is the limiting factor to algae growth, you should in theory be able to control the blooms by reducing discharges of that nutrient.

The lineup of suspects at the time included nitrogen, carbon, potassium, and phosphorus. All of these elements could be found among the multitude of waste streams tumbling into the lake, but biologists of the time had their eyes on one culprit in particular. Water samples showed that the amount of dissolved phosphorus in Lake Erie had nearly tripled between 1942 and 1967, which corresponded with the explosion in algae blooms not only in Lake Erie but also in waterways across the continent. It also matched the period when phosphorus-rich synthetic detergents flooded into the market.

By the late 1960s, the United States was churning out some four billion pounds of detergent annually, and sanitary officials calculated that as much as 70 percent of the phosphorus in wastewater could be traced back to all those boxes of detergent powder in all those basements across the United States and Canada.

Many consumers at the time figured soupy green waters were a fair price to pay for cleaner clothes, but this, it turned out, was a false choice. The primary function of phosphorus in detergent is to neutralize hard-water minerals, thus allowing the workhorse detergent molecule to do its job. A box of phosphorus detergent at the time was, essentially, a box full of water softener. P&G's Tide was nearly 50 percent phosphate by weight; Colgate-Palmolive's Axion was more than 63 percent phosphate; P&G's Biz was almost 74 percent. But by the 1960s, more than a third of US residents in the largest one hundred cities were already served by soft water. This meant the phosphorus in the detergents used by tens of millions of Americans was basically useless as a cleaning booster.

Yet when the detergent makers were pressed to at least label their products' phosphorus content so conscientious consumers might choose a low-phosphorus alternative, they argued the labeling would have the opposite of the intended effect. Their point was the soap operas had done their job too well.

"From some surveys that we have done, and some other information we have been able to obtain, it is our *complete* conviction that the average housewife seeing a higher percent of [phosphorus] content will *automatically* equate this to better cleaning (emphasis supplied)," a detergent industry spokesman testified at a 1969 congressional hearing.

Once again, it was Congressman Henry Reuss who pushed the industry to seek a replacement for a troublesome cleaning ingredient, but this time the detergent industry had no intention of backing down. It had the money and resources (P&G was the top television advertiser at the time) to wage a public relations campaign that argued that modern society could no longer function without phosphorus-packed detergents.

"The picture presented by the Soap and Detergent Association is indeed frightening. America is presented not merely with the choice of clean shirts or clean waters, but rather, if the [detergent] industry position is valid, the necessity of keeping phosphorus in detergents as the barrier between American health and pestilence," the US House Conservation and Natural Resource Subcommittee wrote in a 1970 report. "Eutrophied lakes may be a small price to pay for such a remarkable chemical."

The argument held sway with some of the very government regulators who were supposed to be policing the detergent industry. One federal official even proposed a plan to keep phosphorus detergents on store shelves across the country by upgrading US sewage treatment plants so their equipment could remove detergent phosphorus before it was discharged into rivers and lakes. The cost for the proposed nationwide sewage treatment upgrade was estimated at some $260 billion in today's dollars.

Congressman Reuss pointed out the absurdity of the strategy during the 1969 congressional hearing.

"By and large the phosphate which shows up at sewage disposal plants comes from two main sources—household detergents and human waste?" Reuss asked the assistant secretary of the Interior who supported the plan favorable to industry.

"Yes, sir," the assistant secretary replied.

"And household detergents are made by three major manufacturers?" Reuss pressed.

"That is correct."

"And human waste is made by a couple hundred million manufacturers; is that correct?" Reuss replied.

"Yes, sir."

"Well," Reuss said, "doesn't it occur to you that it is easier to do something about three than about a couple hundred million?"

The detergent industry, meanwhile, argued there was "no evidence" that there was any relationship between detergent phosphorus and the explosion of algae in US waters stretching from the Potomac River to Lake Erie to Lake Washington in the Pacific Northwest, and the galaxy of lakes and rivers in between.

Science was about to provide that evidence. In the biggest way imaginable.

WHEN A TEENAGE David Schindler wasn't working on the family farm in western Minnesota in the 1950s, or slinging one-hundred-pound sacks of potatoes in his grandfather's nearby warehouse, he'd ride his one-speed bicycle around the neighborhood and spend afternoons with his many crushes. Among them were Sallie, Maud, Eunice, Melissa, and Lizzie. These weren't classmates and they weren't members of his church. They were a clutch of sister lakes near Schindler's home about thirty miles southeast of Fargo.

Schindler didn't have the words for them at the time, but it's likely the lakes of his youth would have been classified as mesotrophic—not too shallow, warm, and algae-filled (eutrophic), and not too cold, deep, and relatively fishless (oligotrophic). They were mesotrophic—right in the middle. They were, as Schindler remembers, perfect. Particularly the two-mile-wide, tree-rimmed Lizzie.

A wrinkled Norwegian farmer who lived next to Lizzie had a side business renting boats for 50 cents to young anglers like Schindler. "They were wood-stripped, and they leaked, particularly in the spring before the wood swelled up," Schindler told me. "You could go out and catch your limit in walleyes in a couple of hours just sitting in this lake all by yourself."

But the sister lakes started to age fast as farmers began selling off lakefront parcels for as little as 20 cents per shoreline foot, and cottages began to sprout. Along with those summer homes came holding tanks to contain all the human waste new to the neighborhood. The tanks didn't do a very good job. Schindler said some people even bragged they just used a couple of old barrels or rusted out car bodies as tanks. "At the time," Schindler said, "these systems were very poorly policed."

In just a few years, algae blooms started to taint the lakes, but as the teenaged Schindler headed off for the University of Minnesota he did not understand the connection between what was happening on the land and why the pristine lakes of his youth had so suddenly started to turn soupy.

Schindler did not plan to look back when he headed to the Twin Cities with designs on a career in engineering or physics, but he soon found himself miserable on the urban campus. "I felt trapped," he said. "I couldn't get out of town on a one-speed bike and I was just very unhappy."

His career track bent dramatically while visiting a high school friend at what is now North Dakota State University, or Moo-U, as Schindler called it at the time. Schindler, then a college sophomore on winter break, was killing time in a hallway waiting for his friend's class to end when he struck up a conversation with a professor who mentioned that he had just received a new piece of equipment called a calorimeter, a supremely sensitive device used to trace the flow of heat—energy—as it moves between organisms.

When Schindler mentioned he had done calorimetry work as a student at the University of Minnesota, the professor asked if he might be able to help him with some experiments, and the next summer Schindler had a job in the North Dakota professor's lab painstakingly measuring heat exchange between organisms at infinitesimal levels.

The job involved lots of waiting around between calorimeter measurements, and Schindler filled the downtime devouring the books on the professor's shelf, including *The Ecology of Invasions by Plants and Animals* by renowned Oxford scientist Charles Elton, a book that hooked him just like he hooked all those walleyes back on Lake Lizzie. It inspired Schindler to transfer to North Dakota State, to switch his major from physics to zoology, and to charge straight ahead into the emerging field of ecology.

Schindler thrived at his new university, in the classroom and in his student job in the laboratory with the calorimeter. He also held his own on the football field as a 195-pound defensive lineman for the Bison. By the beginning of his senior year, Schindler had so excelled in every aspect of his college career that the professor he worked for suggested he apply to graduate school at Yale, or maybe Duke. Schindler was more ambitious. He wanted to win a Rhodes Scholarship and study at the University of Oxford under the eminent Charles Elton himself.

Schindler knew it was a long shot, for him or anyone. But it was the only way he could continue with his education; he was so broke at the time that he couldn't afford meals on the train that took him and other Rhodes Scholar hopefuls out to Portland, Oregon, in late 1961 for their interviews. "I sat there miserable and hungry," he said of the train ride, "listening to the others recite Shakespeare and generally have a good time."

The morning of the interview a nervous Schindler spent his last dollar on a hamburger he had to choke down because of nerves and then opened the door to the interview room to make his pitch. The knots in his stomach only tightened when the scholarship committee started to pepper him with unexpected questions, including one that *was* actually related to Shakespeare, about the relationship dynamic between Othello and the conniving Iago.

He doesn't remember exactly how he tried to answer it, but

he did try. Then came a barrage of questions about art history and art theory and it quickly became clear to everyone in the room that Schindler had no clue what he was talking about. One interviewer finally blurted: "You want to go to Oxford to study art? So how come you know so little about it?"

Schindler at this point was more baffled than embarrassed. Then he remembered he had written in his Rhodes Scholar application about his plan to study limnology at Oxford. "Being good scholars," he told me, "they had their Latin dictionaries, and they must have figured my interest was in art because the root of the word—limn—means to draw or sketch. Then the lights went on."

He politely informed the committee that the root word for the discipline he hoped to pursue at Oxford was, ahem, actually Greek in origin. "I explained limnology was the study of fresh-water," he recalled. "And then they just sat there and let me talk for the next 20 minutes."

News that Schindler was among the thirty-two US college students to win a Rhodes Scholarship made the front page of the local section of the Minneapolis *Star Tribune* on Christmas Eve, 1961, and by the next summer he had liquidated his assets—fishing rods, shotguns, and an exhaust-belching outboard motor—to buy a one-way ticket on a ship to England. The Minnesota farm boy was bound for the Oxford laboratory of Charles Elton.

Four years later Schindler was back in North America with a PhD in ecology and looking for a career beyond the laboratory walls he found too confining, both for his outdoor lifestyle—and for the type of experiments he wanted to conduct.

Schindler had become convinced during his time at Oxford that the key to understanding the way energy flows through a lake is not through laboratory gear or "fancy" mathematical models hatched on a chalkboard. He saw lakes themselves as potential laboratories upon which whole ecosystem experiments could be conducted. One problem, of course, is the idea that someone would take a perfectly good lake and dose it with various contaminants and then sit back and watch what happens. And it was impossible to think someone could be given the green light to conduct such experiments on dozens of lakes.

At least it was before Schindler's time.

WHEN SCHINDLER RETURNED to the United States from Oxford, he interviewed for faculty positions at Yale and the University of Michigan, but he found their campuses too lab-focused and too far from the waters and woods that had tugged him into the field of ecology. He instead took a job at Ontario's brand new Trent University, which at the time lacked the academic luster of the other institutions but would allow him to launch a career conducting research on all its nearby lakes and forests. Then, barely a year after he started the job, came a plum offer.

The Canadian federal government and the Province of Ontario had agreed to carve out a swath of wilderness about two hundred miles southeast of Winnipeg, and they were looking for ecologists to conduct lake-wide experiments on the cause of the algae outbreaks on Lake Erie and similarly algae-plagued lakes across the continent. Nearly five hundred lakes were set aside as candidates for "whole-lake experiments" after extensive helicopter surveys of the publicly owned land during 1967. The chosen ones would be treated like oversized lab rats.

The ecologists would be free to manipulate their research lakes in almost any manner they thought would help solve the problem. It was the exact scale of outdoor laboratory experimentation Schindler yearned for, but he reluctantly declined a job when the lab director told him that the conditions would be so rugged he could not bring his wife and young daughter.

Schindler's would-be boss had a point. The area picked to do the open-lake experiments is so remote and so freckled with lakes that when paddling or hiking through it your mind can play tricks. It can start to feel like islands dotting a grand lake, and not lakes dotting a vast forest. And at the time there was also no electricity at the planned research site in the middle of the woods.

The revolutionary outdoor lab remained little more than a sketch on the map when the Canadian government came calling on Schindler the next year. This time he was told he could bring his wife and daughter, along with a newborn son. The catch: the family could not actually live at the research camp where Schindler, still in his twenties, was to be the leader of more than a dozen field biologists recruited from around the world.

Schindler jumped at the second chance. He arrived in spring 1968 and promptly set up a homestead on a tiny island about a five-minute canoe ride from the research station's tents and rumbling power generator. His rationale for living on the island was that he would not have to cut a trail through a tangle of birch and pine trees to get to and from his work. And when he was in the field, sometimes for days on end, his wife would not have to fret so much about dangerous run-ins with wildlife, black bears in particular. The family lived in a red 10- by 14-foot tent on the tiny island that summer. A crib for their baby boy was fashioned from a wooden box that had contained the lab's depth finder.

Colleagues were immediately stunned by Schindler's focus and stamina. "It always amazed me, seeing him come in for breakfast, go back to the lab, come in for lunch, go back to the lab, then come in for dinner, and then back to the lab. Then he'd come out of the lab with his briefcase, hop in his boat and paddle home across the lake," one former colleague recalled. "He told me once he only slept about three hours per night. He

was just fully committed to the project and when you see someone like that you felt guilty if you weren't committed too."

The biologists' first order of business was to collect data on
individual lakes' chemistry, temperature, depths, and aquatic
life. Schindler and his crew did the work by canoe and, when
necessary, by helicopter. That often required Schindler to climb
onto one of the aircraft's pontoon floats to scoop up water samples. The copter's engine would blast him with so much exhaust
it sometimes made him vomit.

By the end of that first summer, the scientists at the outpost
known simply as the Experimental Lakes Area had identified
several dozen lakes to formally set aside for their ecological
investigations.

Once the actual work was ready to begin, Schindler had a
simple plan. The detergent industry had argued that Lake Erie's
algae problem wasn't rooted in phosphorus-rich effluent from
its products but instead caused by carbon-rich household sewage, which of course wasn't the industry's problem. Schindler
decided to put carbon on trial.

"So we had this lake, Lake 227," he said. "And my proposal
was: Let's do an experiment on it that will give a yea or a nay to
this carbon theory. So I said: 'We'll add nitrogen and phosphorus to the whole lake. If we then get an algal bloom this whole
carbon theory will be knocked into a cocked hat.'"

The experiment started the next spring, 1969, on the twelve-
acre lake that was a two-portage, five-mile paddle from the base
camp, which itself was so remote a "road" had to be blasted across
the forested, rocky landscape to bring in the trailers, tents, power
generator, and lab gear. (The road remains so rough today that a
visitor would be foolish not to pay for the extra rental car insur-
ance. Absolutely foolish.)

Even getting a research boat to Lake 227 was a production; it
had to be lashed between the pontoons of the helicopter.

One of Schindler's colleagues still remembers the day they went to work on that first whole-lake experiment. They ripped the starter cord on their fiberglass boat's ten-horsepower engine and motored out to the middle of the lake. Then they cut the engine and pulled the drain plug at the back of the boat. When a plug like that is pulled on an idled boat, it isn't a drain. It's just a hole in the bottom of the boat. Water gushed in.

With lake water fast lapping at their ankles, they put the drain plug back in and dumped in a couple of bags of store-bought phosphorus- and nitrogen-based fertilizer and swizzled the giant nutrient cocktail with a paddle. Then, as they motored off in the water-laden dinghy, they pulled the drain plug again and did loops around the lake to allow the fertilizer-rich water in the boat bottom to flush out the drain hole. It didn't take long to get a verdict on the carbon question.

"We knew the answer two weeks later," said Schindler. "We had an algal bloom."

Since the scientists had added zero carbon to the water, human-introduced carbon was eliminated as the cause for the freshwater algae blooms. Even so, the ecologists calculated that there should not have been enough carbon naturally in the water to produce a bloom as large as the one that appeared after they added the nitrogen and phosphorus.

"The big fascination for us was figuring out where the algae got enough carbon to do that," Schindler said of the Lake 227 results. The researchers, taking around-the-clock carbon measurements, eventually determined that the algae would consume the lake's available carbon during the day while they photosynthesized and grew.

Then, when the sun dipped and the algae shut down photosynthesizing for the night, the carbon-depleted lake itself would essentially take a deep breath of carbon dioxide from the atmosphere to regain its chemical balance. Tests showed that each morning a fresh, nature-supplied dose of carbon was in the water, just in time for the sun to rise and the algae to get back to work growing. This was precisely the reason researchers had turned to the large-scale experiments; laboratory trials couldn't possibly replicate a whole lake's astonishing ability to do something like take a deep breath of carbon dioxide at night.

As word of the experiment's results spread, Schindler said the detergent industry tried to switch its blame for the algae outbreaks away from carbon.

"Although the results of the Lake 227 experiment silenced those who believed that carbon could limit eutrophication, a soap and detergent industry heavily dependent on high-phosphorus detergents continued to argue that phosphorus control alone would not solve the problem," Schindler recalled. "They proposed that nitrogen must be controlled as well because bottle-scale experiments in many lakes showed nitrogen to be limiting for all or part of the year."

So Schindler decided to put nitrogen on trial next. His idea was to take a peanut-shaped lake—Lake 226—and cut it in half by installing a polyurethane divider at the lake's midpoint running from shore to shore, and from surface to bottom. The researchers stitched together their oversized shower curtain out of material designed to corral oil spills and suspended it from the lake surface on a floating line. Divers sealed the curtain on the lake bottom by tucking it underneath a row of heavy rocks.

Once the two lobes had been severed, one lake suddenly became two lakes. Both sides got dosed with carbon and nitrogen. But one side also got fed phosphorus. Once again, just weeks after nutrient dosing began, one side of the lake turned a telltale bright green. It was the side fed phosphorus.

"One of the technicians came in from a helicopter survey

one day and said: Wow! You should see that lake," Schindler told me. "So we went up with cameras and took the pictures that became famous."

That contrast between the pristine, deep-blue water on one side of the curtain and the golf course-green water on the other delivered a knock-out blow to the detergent industry's argument that its phosphorus did not cause the algae problem.

Phosphorus, of course, was not the *only* nutrient needed to produce algae blooms on a lake, but the Lake 226 experiment and subsequent experiments convinced Schindler's team it was the *limiting factor* in algae growth on most if not all freshwater bodies. There are, today, a school of researchers who contend that, on some lakes, nitrogen can be the limiting factor, a notion Schindler scoffed at until the day he died.

Reduce phosphorus, he argued, and algae outbreaks are reduced. Demonstrating this phenomenon to various state legislatures with bar charts showing micrograms of phosphorus per liter on different lakes and the resulting algae densities was one thing. The picture from the helicopter was something completely different.

"That is exactly what the hearings panels needed, because a lot of panel members were non-scientists and you could see their eyes glaze over as we walked them through all the data," said Schindler. "If a picture is worth a thousand words, in science it is probably worth one hundred thousand words. It was just tremendously effective in explaining the role of phosphorus."

EVEN BEFORE SCHINDLER had his picture proof, the public started to become suspicious of the detergent industry argument that the billions of pounds of chemicals it had unleashed on public waters didn't cause the algae problem. Some of the most ferocious criticism came from unexpected places.

"Lake Erie is a disaster. Lake Tahoe is in danger. Puget Sound is an ecological nightmare. The Charles River is a disgrace," proclaimed one advertisement that ran in newspapers across the country in 1970. "There is no more time left. Either we begin to solve our polluting problems *now*, or they may never be solved. We may live out the rest of our lives in a world of garbage."

The half-page advertisement went on to single out detergent phosphorus as the root of the problem. The ad did not come from an environmental organization like the Sierra Club or any other activist group. It came from a *Tide* rival that began pushing its own low-phosphorus detergent formula as the environmentally responsible choice. The ad not only named names as far as brands with the highest levels of phosphorus, but it also listed the percentage of phosphate in those competing brands.

Later that year the US detergent industry agreed to limit phosphorus content in its products to no more 8.7 percent by weight. That was not good enough for Chicago, which ordered an outright ban on phosphorus detergents later that year, a move P&G tried, unsuccessfully, to fight in federal court. Indiana followed with the nation's first statewide phosphorus detergent ban. That measure also led to an unsuccessful lawsuit by the detergent industry, as did similar phosphorus bans in Detroit and Akron, Ohio.

Enough states followed Indiana's lead after 1973 that by the mid-1990s the detergent industry voluntarily pulled phosphorus from household detergents. Detergent phosphorus builders were replaced with, among other things, sodium carbonate. That compound is still used today in powder formulas of Tide, which remains, more than seventy-five years after it was introduced, the dominant detergent on the market.

Similar phase-outs happened subsequently with phosphorus dishwasher detergent. The flow of sewage-based phosphorus waste into public waters was further diminished beginning in the 1970s with billions of dollars in sewage treatment plant upgrades across the country.

And, just as Schindler and other scientists of the time had predicted, North America's algae-ravaged lakes and rivers began recovering throughout the 1980s. Lake Erie bounced back so quickly and so fast that in the mid-1980s Dr. Seuss agreed to remove any reference to Lake Erie in future editions of *The Lorax.*

TODAY THE ALGAE BLOOMS on Lake Erie are back and they are as bad as they were in the dark days of the 1960s dead zones, even worse. This time the decaying algae isn't just sucking the oxygen out of Lake Erie. It is poisoning it. And just like in the 1970s, these blooms aren't confined to Lake Erie but are increasingly plaguing lakes and rivers from Florida to the Pacific Northwest.

Once again the problem is phosphorus. Once again biologists have identified the industry responsible for it.

And once again that industry is being allowed to pollute with impunity.

# CHAPTER 6

# *Toxic Water*

L AKE ERIE MADE its indelible mark on Sandy Bihn when she was a girl living in Toledo, Ohio, in the 1950s, particularly the summer version of Lake Erie. Her father managed a dry-cleaning business that he would shut down every Fourth of July holiday so the family could rent a cottage on the beach just across the state line in Michigan.

Her dad would take the first week off, and then make the half-hour drive back to work every day during the second week, so Bihn and her little sister and their summertime friends could savor their prolonged and mostly unsupervised respite from the steamy city. In the mornings, the girls angled for yellow perch in their rented dinghy. In the afternoons, they swam and romped on the beach. At night they breaded and fried in butter the fish they'd caught earlier in the day.

Some days, when the water was too rough after northeast winds kicked in, as they always seemed to do for at least a few days straight each vacation, Bihn and her sister would comb the shoreline for pretty stones or go for hikes inland. Whatever she did,

Bihn did it barefoot, and by the end of every vacation the soles of her feet had turned so leathery she could gambol down the gravel roadway shoulders near their cabin without a wince. "It was just my favorite time of year," Bihn says, "and I knew from early on that no matter what I did in my life, I needed to be by that lake."

In 1987, as Lake Erie was well on its way to recovering from the detergent-driven Dead-Sea days of the 1960s and '70s, Bihn and her husband built a house on the lake's western shoreline near Toledo. The lake's water quality had improved so much by that point that Bihn never felt the need to hop in a car to take a vacation; her kids spent their own barefooted summer days on the beach in their own backyard, swimming off an anchored raft they called "The Island."

Today The Island is gone. So are the deep-blue waters that marked the lake's resurgence in the 1980s and 1990s. Lake Erie, once again, suffers such chronic algae blooms that Bihn can't remember the last time anyone in her family kicked off their shoes to dip so much as a toe in the surf. Any swimming the family does today, in fact, is confined to the chlorinated waters of a small backyard pool the family built after the lake's soupy water gave Bihn's husband one-too-many ear infections. The distressing decline of Lake Erie has put Bihn on a mission.

"All I want," Bihn told me, sitting in her living room that is as nautically appointed as a 1980s' Red Lobster lobby, "is for that lake not to be green."

The green she is talking about isn't a pickle-juice green. It's an emerald green, thick as a gallon of paint. And it is toxic.


LAKE ERIE'S MODERN algae plague can't be pinned on detergent makers, industrial dumping, or sewage treatment plant discharge. All these pollution sources are heavily regulated by modern pollution discharge rules. But this is not the case with the farming industry, which is fueling today's blooms. Specifically, the lake's algae troubles can be traced to excess phosphorus fertilizer washing off croplands in the vast, flat, and fertile Maumee River basin on the lake's western end.

Prior to white settlement, the Maumee region was known as the Great Black Swamp, a roughly 1,500-square-mile morass, rich with wildlife, that functioned for thousands of years as a natural filter for turbid, naturally nutrient-rich rainwater runoff bound for Lake Erie. White settlers drained and tamed the swamp with a system of ditches and underground pipes beginning in the 1800s, and today the once-soggy Maumee River basin is home to some three million acres of laser-straight rows of corn and soybeans (along with lesser amounts of wheat, hay, and oats) and a ballooning number of livestock operations. Once part of an immense water purification system, the Maumee River now functions more like a syringe that each year mainlines thousands of tons of excess agricultural phosphorus directly into Lake Erie. This is the fuel driving Lake Erie's modern, massive algae blooms, even if farmers are quick to deflect responsibility for them.

They say the problem can be traced back to dirty sewage treatment plant discharges and industrial pollution as well as over-fertilized golf courses, lawns, and even homeowners' leaky septic tanks. They're not altogether wrong, but those sources combined account for only 15 percent of the annual phosphorus load carried into the lake by the Maumee River. Agriculture is responsible for the rest, even though biologists trying to solve the lake's phosphorus problem acknowledge that farmers in the Maumee watershed, on the whole, have in recent years actually been reducing the amount of factory-made chemical nutrients they apply on their crops.

Farmers are also working with government regulators—and the public's money—to keep the chemical fertilizer that they do use from washing off the landscape before their crops can take it up. There is government money to pay farmers to plant "cover crops" after the summer growing season to soak up excess fertilizer and anchor phosphorus-saturated soils so they don't wash into the lake. Government funds also subsidize farmers who plant buffer zones of vegetation at the edges of their fields to capture fugitive phosphorus particles before they hit the ditches and rivulets that flow toward Lake Erie. Another publicly funded program gives farmers money to install gates to slow flows of phosphorus-contaminated stormwater in the underground pipes that drain farm fields of standing water.

Yet the blooms rage on.

Part of the problem is "legacy" phosphorus leaching from fields overdosed during decades when chemical fertilizer was relatively cheap. Another factor is climate change—increasingly intense spring rains can wash away the fertilizer farmers spread before it can be taken up by the year's crops. This has become a particularly acute problem in recent years as farmers have turned to "no-till" farming that leaves fields smooth and level as a parking lot. The practice protects topsoil from washing away but it leaves the crusty layer of fertilizer spread each fall primed to melt and wash into the lake when the April showers hit.

But Bihn, executive director of the environmental group Lake Erie Water Keeper, puts most of the blame for the lake's phosphorus overload on the exploding numbers of livestock inside the Maumee watershed.

"It's not that commercial fertilizer isn't part of the problem, but manure has been swept under the rug," she says. "It's crazy, what they are getting away with."


LAKE ERIE'S TWENTIETH-CENTURY pollution troubles led to bans on phosphorus detergents and prodded Congress to pass the 1972 Clean Water Act, which required cities and industries to dramatically ratchet down fertilizer and other pollution discharges into rivers, lakes, and coastal waters across the nation. But the landmark environmental law largely gave agriculture a pass.

The rationale at the time was that the necessary nutrient reductions to heal US waters could be achieved by pulling phosphorus from detergents and by reducing nutrient-rich discharges from industries and cities. Not only were these the biggest source of phosphorus pollution, but regulating them was also fairly easy compared to trying to control the diffuse (in regulatory parlance, "nonpoint") chemical fertilizer and manurial pollution dribbling off the agricultural landscape. You can catch and treat the "point source" pollution carried in pipes. Getting a handle on pollution (excess nutrients) spread across millions of acres of open land is another matter.

But in the half century since the Clean Water Act was passed, America's farms have undergone such dramatic changes that they've become a lot more like the point-source polluters.

American agriculture today is often conducted on an industrial scale. Just like any factory, farms that can have more than ten thousand head of cattle produce a predictable daily load of pollution (manure) that is anything but diffuse. Farmers liquefy the stuff and pump it into pond-sized sewage lagoons that can hold millions of gallons. To keep those lagoons from overflowing, farmers must regularly spread their phosphorus-rich waste on farm fields—sometimes even if those fields do not need a nutrient boost.

The biggest of these factory farms, referred to as concentrated animal feeding operations (CAFOs) by regulators, are required to get permits that regulate how manure is managed in areas where it is concentrated on a farm, places like barns and manure lagoons. But those permits are often loosely enforced, and the regulating largely stops once that manure is trucked off the farm and spread on nearby pastures. And farmers who want to prevent government regulators from even knowing how much manure they generate, and how and where they dispose of it, can do so by keeping their farms under a certain size, even if they're still mammoth.

An Ohio pig operation with fewer than 2,500 hogs, for example, or a fowl farm with fewer than 82,000 egg-laying chickens, or a dairy farm with fewer than 700 cows, can qualify to keep its manure disposal operations largely unregulated and out of the view of the public.

But a picture of the extent of the Maumee watershed's growing manure burden began to emerge after conservationists got around the secrecy by using aerial photos in a 2019 analysis of recent farm expansions.

There are government standards for how much indoor space is required for each type of farm animal (a mature dairy cow requires an average of 80 square feet; a pig needs 7.5 square feet; an egg-laying hen demands 67 square inches). So by analyzing photographs of newly built and expanded farm structures that have distinct profiles depending on the animals they house, the study authors worked backward from a barn's size and shape to calculate how many and what type of animals were likely inside. It's not a perfect survey, but the conservationists contend their estimate of livestock numbers is better than anything compiled by anyone so far, government regulators included.

Their findings are stunning. The number of animals in the Maumee watershed more than doubled to twenty million between 2005 and 2018 while the amount of manure-based phosphorus added to the watershed grew by 67 percent, to 10,600 tons annually.

Those farming operations collectively produce as much excreta as a city of at least several million people. But the difference is city sewage treatment plants pull significant amounts of the bad stuff from the human waste that they process, phosphorus included. The Maumee's manure does not go through wastewater treatment to remove its chemical and biological pollutants. Instead it gets spread on croplands to make room for the next day's batch of poop. Little if any of the manure produced each day gets exported outside the watershed because of the expense of trucking such massive amounts of slop; economics dictate manure typically doesn't get spread on fields more than about ten miles from the animal that produced it.

And whatever doesn't get taken up by a kernel of corn, a soybean, or a stalk of wheat does what everything else does in this world, even in the almost flat-as-a-plate Maumee watershed—it flows downward, and in this case that means it eventually winds up in Lake Erie.

Ohio farmer Bill Myers grows corn, soybeans, and wheat across the street from Maumee Bay State Park on the western shore of Lake Erie, on the same land his great-grandparents farmed after arriving from Germany in the late 1800s. He says he doesn't use animal waste as fertilizer because his crops are too far from any of the big manure producers. He does say he'd use the stuff if he had a cost-effective source, because it not only provides ample amounts of fertilizer but also contains organic matter that keeps soils healthy. But he acknowledges some percent of the manure generated in the Maumee watershed—he doesn't pretend to know how much—is spread on fields, even if they don't need the organic matter and nutrient boost, because Maumee basin farmers would be drowning in manure if they didn't regularly unload it on the landscape.

"They're trying," Myers told me, casting playful conspiratorial glances from side to side, "to get rid of their shit."

Myers is as frustrated by this as anyone, but he dismisses the idea that fellow farmers are trying to get rich by dumping their waste on the cheap. He says most of them are just trying to survive and that the general public doesn't realize how economically challenging it has become to farm in the twenty-first century. His fourteen-hour days earn him an annual income, he estimates, somewhere between $35,000 and $50,000.

"Of course we need to make a living. We don't do this for free," he told me, a pinch of tobacco bulging from his lower lip. "Would you walk around with that notebook in your hand talking to all sorts of people for free, just because you're a generous guy? Everybody needs to get paid."

Myers said he knows plenty of farmers who recognize their role in Lake Erie's phosphorus problem and are doing what they can to fix it. But he also acknowledges the existence of a super-polluting minority, speculating that about 20 percent of the farmers in the basin do about 95 percent of the manure polluting. He can understand why people like Bihn are pushing politicians to better regulate how farmers manage their manure because the situation is becoming increasingly unacceptable.

"There's always going to be times when we're not happy about the algae in the lake. If we get it down to one in ten years, people can live with that," he said. "If it's happening nine out of ten years, well then people are going to be pissed. And people are pissed right now."

LAKE ERIE'S MID-TWENTIETH-CENTURY algae troubles faded not only because phosphorus was pulled from detergent formulas. Some $8 billion was also spent on sewage treatment upgrades that allowed cities across the Midwest to slash the amount of sewage-based nutrients they discharged into Lake Erie and the other Great Lakes. The upgrades were part of the US and Canadian recovery plan for Lake Erie that called for cutting the amount of phosphorus allowed to flow into the lake from an average of twenty-nine thousand tons a year to eleven thousand tons—a number regulators calculated would solve the algae problem, and they were right.

Today the total annual phosphorus flow into the lake remains below that eleven-thousand-ton threshold. But the blooms are back and big as ever, largely because much of the phosphorus now washing off agricultural lands hits the lake in a super-potent dissolved form.

Another change since the 1960s is that those earlier algae blooms were made up of an assemblage of species, most of them nontoxic. Today's blooms are often caused by "blue-green" algae, colonies of which are now ravaging Lake Erie—and rivers and lakes across the United States.

Blue-green algae aren't technically algae but a type of photo-synthesizing bacteria. Also known as cyanobacteria, these single-celled organisms can produce a liver toxin potent enough to kill dogs and make a swimmer vomit in a matter of seconds after an accidental gulp.

Cyanobacteria may be toxic, but they are also completely natural. Fossil records show cyanobacteria have been thriving on Earth for at least 3.5 billion years (the oldest rocks are roughly 4 billion years old), and this is a good thing. Cyanobacteria literally breathed modern life into the planet; their collective respiration is what put enough oxygen into the atmosphere some 2 billion years ago to open the door for life on Earth as we know it today, including human life.

But various species of phosphorus-loving cyanobacteria also produce plumes of potent toxins, a phenomenon that has been recognized for more than a century. In an article that appeared in the journal *Nature* in 1878, an Australian chemist reported a lake near the mouth of the Murray River with a "scum like green oil paint" and "thick and pasty as porridge." The livestock he observed lapping up the contaminated water quickly descended into stupor before convulsing and collapsing. The chemist then fed the poisonous water to various other species and found different animals took different periods of time before the toxicant took its toll. "Time—sheep, from one to six or eight hours; horses, eight to twenty-four hours; dogs, four to five hours; pigs, three or four hours." No matter how long it took, the end result of extreme exposure to the toxic algae was the same: death.

South African researchers similarly reported in the early twentieth century "many thousands" of cattle and sheep dying from algae-contaminated waters, along with horses, mules, donkeys, hares, poultry, and waterfowl. Necropsies on livestock killed after being fed water from a South African reservoir contaminated with blue-green algae showed liver damage so severe that their blood-purifying organs had turned black as coal.

Zebra and quagga mussels, native to the Caspian Sea region, are implicated in this new wave of toxic algae blooms because the diminutive filter feeders that carpet the lake bottom gobble up almost anything floating in the water—except the blue-green algae, which are consequently largely left alone to thrive with little competition. So when a mussel-infested lake, even a Great Lake like Erie, gets an algae bloom these days, it's more likely now to be of the toxic sort.

Blobs of this cyanobacteria can now sprawl across some two thousand square miles of Lake Erie during an exceptionally bad algae year, and as farmer Myers noted, in the last decade almost all years have been bad.

The outbreaks are more than a problem for biologists, swimmers, and fishermen. In August 2014, a plume of toxin produced by cyanobacteria got sucked into the Lake Erie water intake operated by the Toledo public water system and forced officials to issue a do-not-drink order in the middle of the night. Boiling the contaminated water did not make it safe to drink, health officials warned. It actually concentrated the toxin.

Word of the order spread so quickly and so rattled Toledoans thatjust several hours after it was issued, stores up to an hour away were out of bottled water. As panic mounted, National Guard troops from all corners of Ohio mobilized to bring in truckloads of bottled water, portable water treatment systems, and pallets of baby formula to sustain residents until their water supply could be made safe through additional chemical treatment.

Two months after the crisis, mayors from more than a dozen Great Lakes cities, all of which depend on the lakes for their drinking water, met in Chicago and vowed to take whatever steps were necessary to avoid a repeat of the Toledo debacle anywhere else on the Great Lakes—a drinking water source for more than thirty million people.

It took four years, but in 2018 the US and Canadian governments finally reached an agreement to slash the annual phosphorus flow into Lake Erie by 40 percent by 2025.

Scientists are confident that such a nutrient diet will cure Lake Erie, just like the phosphorus reductions of a half century ago did. But this time it isn't happening because this plan doesn't come with any new laws to force the reductions. All the cutbacks are essentially voluntary, despite outcry from almost everyone but the farmers and the politicians afraid to challenge them.

"We live in a state where our legislature is a wholly-owned subsidiary of the farm bureau," one critic grumbled in a story that appeared in the *Toledo Blade* in spring 2018. "I'm sorry, but it's true."

This was not the rant of a professional environmentalist.

It came from the mayor of Toledo.


IN MID-SUMMER 2019, I climbed aboard a ferry about sixty miles east of Toledo and, along with more than one hundred scientists, politicians, environmental activists, agriculture experts, reporters, and professional fishing guides, headed for a biological research station on a Lake Erie island. We were bound for what has become a grim annual event for Ohioans—the official forecast for how bad Lake Erie's toxic bloom will be when peak algae season hits in late summer.

It was July 11, typically too early for a budding blob of blue-green algae to appear at the mouth of the Maumee River where it spills into Lake Erie—a key sign that algae season has arrived. But based on the amount of fertilizer farmers reported spreading on their fields for the upcoming growing season, the severity and timing of spring rains, summer water temperatures, and long-term weather patterns, scientists are now able to predict with impressive accuracy how big a summer bloom will get before the toxic gunk dissipates with fall winds and cool temperatures.

It was a glorious morning for a trip to the forested island located about four miles south of the lake's invisible US and Canadian border. The place is freckled with modest cottages in the shadow of a 352-foot-tall (taller than New York Harbor's Statue of Liberty) granite-faced monument to Commodore Oliver Hazard Perry's victory over the British fleet during the War of 1812. The clash prompted Perry to famously declare in a letter to his commander: "We have met the enemy and they are ours."

Perry's quote is almost as famous as the riff on it that appeared in Walt Kelly's beloved newspaper comic strip *Pogo* on Earth Day in 1971, in which the namesake character, an opossum, looks upon a forest floor despoiled with all manner of trash and declares: "We have met the enemy, and he is us."

That happens to be a particularly apt assessment of the modern-day battle for Lake Erie. It can be argued that anyone who eats a bagel, buys ice cream, or feasts at a family barbeque bears some responsibility for the lake's troubles—if you have a problem with what farmers are doing, the argument goes, then don't eat!

But US agriculture today isn't just about putting food on dinner tables. Much of the corn grown on the landscape drained by the Maumee River ends up as ethanol in our gas tanks. What's left over is mostly used to feed cattle or processed into soft-drink sweeteners. Soybeans harvested in the watershed are used for biodiesel and animal feed.

"We're not actually really growing food for humans there right now," says Jennifer Blesh, a University of Michigan agroecologist. "Most of it is technically raw materials for other products."

Yet agriculture in western Ohio isn't just an industry—it's a critical part of the region's identity. And so is Lake Erie.

This tension is in the blood of Democratic Ohio legislator Michael Sheehy, a passenger on the ferry that morning. Sheehy's mother grew up on a farm, and he insisted he has enormous respect for Ohio's largest and most politically influential industry, but he is not afraid to speak out about the harm that industry is causing.

"This lake is one of the most marvelous gifts God gave this planet, and for us to treat it with such disrespect, such disdain, is unconscionable," he told me as we disembarked from the ferry and headed for the press conference, where Ohio agriculture officials reported that less than 50 percent of the typical load of chemical fertilizer had been applied for that growing season because heavy rains had kept farm equipment off the soggy fields. They also reported that just a fraction of the normal amount of manure was spread. It sounded like a dubious claim. Just like humans, cows, pigs, and chickens don't stop pooping just because the weather turns wet, and a farm's manure needs to be regularly spread because there is only so much capacity in in its manure lagoons.

Despite the reported reduction in nutrients spread on farm fields before that growing season, the news was bad. Researchers reported that day that enough phosphorus had still flushed into the lake after the spring's heavy rains to push the predicted bloom to a 7.5 rating on the state's scale of 1 to 10. This meant they expected it to be larger than the 2014 bloom that had poisoned Toledo's drinking water system.

Nobody speaking at the press conference mentioned that state regulators had refused to take even the modest step of declaring the western end of the lake "impaired" under the Clean Water Act until a lawsuit filed by environmentalists forced them to do so the year before. The court ruling required Ohio to put a limit on how much phosphorus is allowed to spill into the lake's western basin annually, though it does not include penalties for violations of that limit. Nor did any of the speakers mention that, despite the court ruling, the state had at that point refused to even set such a limit.

And nobody in the room mentioned the fact that less than a week earlier, the Ohio Department of Agriculture green-lighted a hog farmer's request to double the size of his livestock operation inside the Maumee watershed by adding 4,800 swine capable of producing about one million gallons of manure annually. The operation's stated plan to dispose of that waste: spread it on the land.

Dave Spangler, the late vice president of the Lake Erie Charter Boat Association, could only shake his head on the boat ride back to the mainland. He was careful to ensure I was clear that he supports farming—but not at the expense of his own industry.

"Everything they're doing is basically voluntary, and the lake is just as bad as it ever was," said Spangler. "I don't know what the answer is, other than more regulation. I see no other way."

The night before the conference, I met with its host Chris Winslow, director of Ohio State University's Sea Grant, a federally funded program that applies academic research to fix real-world problems facing public waters. Winslow said the key to Lake Erie's recovery is smarter farming—using the right type of fertilizer, the right amount of it, and spreading it at the right time and in the right place, etc.—and not necessarily more regulation.

"Would you want the government coming onto your property and telling you what you can do with it?" Winslow, a fisheries biologist by training, responded when I asked if it was, finally, time for new laws regulating the farms emitting so much pollution.

I replied: Well, yes, but it's not about what an individual is doing on his private property. It's about the consequences those actions have downstream—on *public* waters.

"Yeah, I get that point," he said, "a tragedy of the commons."

What is happening in Ohio is actually a tragic loss of common sense.

Several weeks later the toxic bloom hit, and it was almost the exact size the scientists predicted, roughly seven hundred square miles—half the size of Long Island.


WHILE LAKE ERIE suffers because Ohio government officials have refused to force even modest steps to make the phosphorus assault on Lake Erie stop, it is a different story about 450 miles to the northwest on Wisconsin's Green Bay. Or at least it should be.

The 120-mile-long arm of western Lake Michigan is something of an ecological twin to the western basin of Lake Erie in that it is shallow, warm, fish-filled—and it, too, is being suffocated by massive algae blooms. Unlike the regulators in Ohio, Wisconsin environmental officials declared the southern portion of Green Bay "impaired" under the Clean Water Act over a decade ago. That required the state to put the phosphorus-overloaded bay on a nutrient diet. It has yet to make much of a difference because the state can make any plan it wants, but it can't do much to force farmers to change their polluting ways, even as farmers continue to ratchet up the size of their operations. A half century ago (at the time the Clean Water Act was passed), a farmer with one hundred cows was considered a big operator. Now some Wisconsin dairy herds have as many as eight thousand cows.

A general guideline for sustainable dairy farming is that each grazing cow requires roughly two to three acres of pasture. It's not an exact number because soil types, weather patterns, and pasture grasses vary. But that much land is basically what is required to not only to generate enough grass to sustain a cow, but also for the land to safely absorb the manure that cow produces. And that manure then fertilizes the pasture's grasses, the cow grazes on them, and then the cow poops some more so more grass can grow. And on and on it goes—a virtuous cycle.

Those days are long gone for so many farms across the country, including in Wisconsin's aptly named Brown County on the southern end of Green Bay. The rapidly suburbanizing county in the heart of "America's Dairyland" is home to some 125,000 head of livestock squeezed onto roughly 190,000 agricultural acres. Most of the Brown County cows today don't graze pastures but are kept in barns and fed farm-raised grain. And each of those cows can produce, by some estimates, eighteen times the amount of fecal waste a human does. None of it gets treated at the Green Bay public sewage treatment plant. Much of it ends up in the bay of Green Bay, where it triggers algae blooms and oxygen-starved dead zones so severe that fish are literally trying to escape from their own waters, Dr. Seuss-style. Homeowners along the shores of Green Bay have actually used leaf blowers to try to push thousands of flopping, asphyxiating fish back into the water.

"How long could you survive with a plastic bag on your head?" a biologist who investigated one of Green Bay's massive fish die-offs once asked me, "because that is what these poor fish are going through."

I grew up in the 1970s less than a mile from the Fox River that flows into Green Bay, and I was forbidden to swim or even play on the riverbanks as a child. It wasn't that my parents were afraid I would drown. It was that they considered the river a liquid dump, and that is what it was. The pollution from paper mills lining the river has since been reined in by the Clean Water Act, and river quality has dramatically improved. But the beaches along the shore of Green Bay are permanently closed to swimming, and phosphorus overloading from Brown County dairies is one of the reasons.

Look at a globe. If you love freshwater, there are fewer places on Earth as attractive as eastern Wisconsin and its nearly five hundred miles of Lake Michigan shoreline. Swimming in open water is not a crazy notion for a lakeside city of 105,000 people, far from it. Kids in cities as big and industrial as Chicago, Los Angeles, and New York City can safely take a dip at their neighborhood beaches. Yet the phosphorus reductions that could help re-open Green Bay's beaches won't happen anytime soon, and it's due to how the Clean Water Act works. Or how it doesn't work.

Green Bay's phosphorus reduction plan under the Clean Water Act called for expensive water treatment upgrades from phosphorus-discharging industries like paper mills as well as the Green Bay metropolitan sewage treatment plant. But, thanks to the Clean Water Act, those phosphorus emitters have already been forced to spend untold millions of dollars on treatment systems to ratchet down their discharges, so much so that it will be hugely expensive to reduce them further.

Yet the Clean Water Act's "nonpoint" exemption for agriculture means Wisconsin regulators cannot require similar phosphorus discharge reductions from the agriculture industry, which is, by far, the largest single source of phosphorus discharged into Green Bay.

Agriculture—chemical fertilizer and manure—is responsible for nearly 50 percent of the phosphorus discharged annually into the Fox River that feeds Green Bay. Factories and sewage treatment plants discharging their treated wastewater into the river are responsible for about 21 percent and 16 percent of the bay's annual phosphorus load, respectively. Much of the remaining phosphorus comes from stormwater runoff. But because the Clean Water Act doesn't include legal authority to compel farmers to change the way they operate, environmental regulators are faced with the prospect of forcing factories and the sewage treatment plant to spend hundreds of millions of dollars to upgrade their already state-of-the-art water treatment systems, a move that may yield little or no environmental benefit.

The Green Bay sewerage district crystalizes the dilemma. It serves about 235,000 residents and sends about twenty-six thousand pounds of phosphorus into the bay of Green Bay each year. That is less than 6 percent of the bay's overall annual phosphorus load.

To meet the state's phosphorus reduction plan, regulators told the district it had to slash its annual phosphorus discharges by about nine thousand pounds. Sewage treatment plant operators say upgrading their treatment system to pull that relatively small amount of phosphorus from its waste stream could cost about $100 million, even though scientists say it might not make a meaningful difference in the pollution levels in Green Bay. It's a similar story with industries discharging into Green Bay that have also vastly upgraded their own water treatment systems in recent decades.

"We could spend $1 billion, and if we're not wise, we could see no water quality improvement," one state regulator told me.

To work around agriculture's loophole in the Clean Water Act, the Green Bay sewerage district tried an experiment with a handful of farms operating in one small Green Bay tributary. The farmers were given grants to undertake farming practices designed to reduce the phosphorus washing off their fields—installing stream buffers, planting cover crops, etc. The sewerage district's idea was to demonstrate that paying farmers not to pollute is far more cost effective than spending hundreds of millions of dollars further ratcheting down industrial and sewage treatment plant discharges. The experiment was a success, and the sewerage district now plans to scale up the program in the watershed.

That makes sense from a regulatory and environmental perspective—spend the money where you can get maximum phosphorus reductions. Others wonder whether it would make more sense—and be more fair—to just change the Clean Water Act so regulators can compel farmers not to pollute.

"I flush my toilet in the city of Green Bay and my rates are going to go up, and it's to subsidize agriculture," one former sewerage district employee told me of the plan to pay farmers not to pollute. "Why is the sewerage district being held responsible for someone else's waste?"

Even though Wisconsin regulators can't force farmers to follow its phosphorus reduction plan, on paper that plan still calls for exceedingly steep cuts in their phosphorus discharges. The goal for just one tiny creek feeding Green Bay is to reduce its annual phosphorus loading from more than thirty-eight thousand pounds per year to just over six thousand pounds. This can't be achieved without a radical reduction in the number of cows allowed on lands straddling that creek, or a dramatic change in how their manure is disposed of.

Yet the farmers responsible for so much of Green Bay's pollution remained unaware of the plan's details, even years after it was adopted.

I once asked the operator of one eight-thousand-cow factory farm how he expected to help meet the reductions called for in the plan. His reply: "You could probably enlighten me more than I could enlighten you."

◊　◊　◊

GREEN BAY AND LAKE ERIE have made headlines for their algae blooms because of their size and the large human populations along their shorelines, but similar stories can be written about lakes and rivers across the country. In fact, you don't have to travel far from Green Bay to see how both ubiquitous and severe the problem has become, even on some of the last lakes you would expect.

A prime example is fifteen-square-mile Lake Mendota on the northern edge of the campus of the University of Wisconsin-Madison. Because the university's renowned center for limnology (the study of freshwaters) is literally built out over its shoreline, Lake Mendota has for decades been one of the most studied lakes in the world. It's also become one of its most algae-ravaged.

The problem, not surprisingly, can be traced back to the dairy operators in the lake's watershed. The landscape has been so saturated with nutrients for so long that scientists say even if phosphorus applications—manurial and chemical—were banned tomorrow, it could be generations before nutrient levels in the watershed's soils dropped to the point that they no longer trigger blooms in the lake.

At the university student union on the lakeshore there is still a swimming dock with a lifeguard chair, but come late summer too often the only life you'll find in the swimming area is guacamole-thick mats of dying blue-green algae.

Undergraduate Camryn Kluetmeier was one of the few people I saw on a steamy August afternoon in 2019 who had actually ventured beyond the lakeshore—not to swim, but to prepare one of the university's research boats to take water samples for blue-green algae. Kluetmeier was twenty years old. She was raised in Madison and spent her childhood summer days splashing on city beaches.

Now, instead of enjoying her lake, she is tending to it like a friend from childhood who has gotten sick. "It's crazy having grown up right here and seeing the changes in my life," she said. "We used to swim all summer. Now, when July hits, swimming is just over. You don't go in. It's just sad."

This isn't only a story just about Sandy Bihn's Lake Erie or Camryn Kluetmeier's Lake Mendota or the Green Bay of my youth. In late 2019, the journal *Nature* reported algae blooms had worsened since the 1980s on nearly 70 percent of water bodies studied in a large lake survey that involved every continent but Antarctica. And a map of US lakes and rivers suffering from blue-green algae outbreaks today looks like, well, a map of the United States.

Hundreds of streams, rivers, ponds, reservoirs, and lakes from Florida to Maine to Washington State to Southern California to Texas to North Dakota and virtually all the states in between are now suffering from similar phosphorus-driven, toxic algae infestations. These annual blooms are already costing the United States more than $4 billion in damage to fisheries, recreation, and drinking water supplies, and scientists say warmer temperatures and increased phosphorus runoff will likely only exacerbate outbreaks across the globe.

Toxic algae blooms are already spreading in ways that baffle biologists. Lake Superior, once believed to be too cool and nutrient-hungry (oligotrophic) to support blooms of cyanobacteria, recently began suffering them. And, for the first time, in 2019 a substantial outbreak of a freshwater type of blue-green algae hit the Gulf of Mexico, a saltwater body that many ecologists thought should be immune to such a bloom.

To find out how this happened, I hopped in my car that hot August day right after I interviewed Kluetmeier in Madison, and headed straight for the next day's opening day of the Iowa State Fair.

# CHAPTER 7

## *Empty Beaches*

Ⅰ HAD A simple question for Joe Biden when he showed up on the first day of the 2019 Iowa State Fair to take his turn atop its famous "soapbox"—a hay bale–festooned stage upon which presidential primary candidates take unscripted questions from everyday fairgoers.

I drove to Des Moines because I wanted to ask Biden what he thinks about corn. Specifically, I wanted to know if he would continue to support the 2005 federal mandate that requires about 10 percent of US automobile fuel be derived from renewable sources, which is mostly corn.

I'd postponed a family vacation to make the trip to Des Moines because I knew Biden would show up. All the serious presidential hopefuls do. The state caucuses have been a bellwether contest for candidates since Iowa moved to the front of the presidential primary calendar a half century ago. This means that what is important to the voters in a small (three million residents), rural (roughly 90 percent of its landmass is farmed), and white (more than 90 percent) state becomes important to every serious presidential contender. This makes Iowans accustomed to fawning soap-box performances.

"We're here for the pandering," one soapbox regular told me on that roasting early August day as we waited for the candidates to arrive. "They are all here to pander, and we love it!"

Minutes later, Biden, decked out in a blue polo shirt and aviator sunglasses, hopped on stage and immediately started to pepper the crowd with platitudes. "We choose unity over division. We choose science over fiction," he hollered. "We choose truth over facts!" (That last line almost made sense in the heat of that day—almost.)

But the truth is the "ethanol mandate," passed under President George W. Bush with overwhelming bipartisan support, is increasingly viewed as a sham—economically, environmentally, and even morally. The idea behind the policy, known more formally as the renewable fuel standard program, is a noble one—wean the United States off its foreign fuel addiction and at the same time reduce greenhouse gas emissions. But it turns out ethanol provides only a slim carbon emissions benefit (and may well offer no benefit at all) because of the energy it takes to plant, fertilize, and harvest the roughly twenty pounds of corn you need to squeeze out a gallon of ethanol. That gallon also contains about one-third less energy than a gallon of gasoline, and it can be notoriously corrosive inside a car engine.

Ethanol's environmental costs soar when you factor in the toll the mandate has taken on the landscape. Less than a decade after the legislation passed, the number of US acres sprouting corn and soybeans (used to make biodiesel, another component of the federal renewable fuel program) grew, by some estimates, more than sixteen million acres—an ocean of kernels and beans larger than Vermont, New Hampshire, and Connecticut combined.

And then there is the moral question of taking food out of

mouths to put it in gas tanks—a staggering 40 percent of the corn grown in the United States is now used to make ethanol.

I had a front-row seat at the soapbox but still failed to catch Biden's eye as he fielded questions during his twenty-minute stump. I did manage to sneak my way behind the stage afterward and got to ask my question when a surprised Biden emerged from a bathroom a couple of minutes later.

He put his hand on my shoulder and said yes, he does support the federal ethanol mandate. But, he added, the key is to pursue "advanced" biofuel technologies that can make fuel not just from corn kernels but also from the fibrous materials in the non-food components of a corn stalk, as well as from other plants. Now, that actually could prove to be a big environmental and economic winner. In the universe of biofuel technology, in fact, development of an economically viable "cellulosic ethanol" is considered the Holy Grail—and it remains today just as mythical and elusive.

*But what about today's ethanol?* I pressed the future president.

"Today," he told me as he spun away, flanked by aides and bound for an ice cream stand, "I support it!"

This was good news for a state in which the renewable fuel industry now supports more than forty thousand jobs. But it was bad news for just about everyone downstream.

Today's freakishly tall stalks of corn, after all, don't just grow on sunlight and water. The government's ethanol mandate is, effectively, also a fertilizer mandate—one that is having increasingly dire environmental impacts in the Mississippi River basin, all the way down to the Gulf of Mexico, where a massive man-made dead zone now rolls in every summer with all the predictability of the tide.

It is the grim consequence of the Mississippi River (along with its little-sister river, the Atchafalaya, that drains away some of the Mississippi's flows through south-central Louisiana) carrying into the Gulf the collective pollution burden of nearly half the continental United States. It includes an annual load of roughly 1.6 million tons of nitrogen and 150,000 tons of phosphorus that stoke a summertime explosion of phytoplankton in the Gulf that, like in Lake Erie, eventually dies, and its decomposition sucks so much oxygen out of the water almost nothing can survive along the seafloor. Biologists call the phenomenon hypoxia, and in some years the Gulf's hypoxic zone can span a Massachusetts-sized eight thousand square miles. This is a not-so-natural disaster for the Gulf's commercial seafood operators who produce about 40 percent of the nation's harvest. They have a problem with Iowa, whether they know it or not.

The landscape leaching the nutrients responsible for the Gulf's dead zone stretches across 40 percent of the continental United States, as far west as the Rocky Mountains and as far east as Pennsylvania. But in the middle of it sits ethanol-addicted Iowa, where nutrient runoff has, by one measure, increased nearly 50 percent in the last two decades and is now such a significant contributor to the Gulf's overall nutrient load that scientists say the state has become *the* battleground to reduce the Gulf's dead zone. "Iowa has a dominant role in this Gulf hypoxia," says one Iowa researcher. "If we solve Iowa, we solve the Gulf."

The algae blooms behind the Gulf's dead zone are fueled by both phosphorus and nitrogen, though nitrogen was the nutrient measured in the study that put Iowa in the crosshairs because it is the driver for the Gulf blooms; it is the element that has been commonly recognized as Liebig's limiting factor for the saltwater coast's explosion in algae.

Phosphorus, therefore, had not been considered the primary problem for the Gulf's algae woes.

At least it wasn't until 2019.

◊　◊　◊

ABOUT THIRTY-THREE SERPENTINE river miles upstream from New Orleans there is a peculiar dam that doesn't run across the Mississippi River but instead parallels its eastern bank. The US Army Corps of Engineers built the concrete and timber palisade in the 1930s after the Great Mississippi Flood of 1927 that left some six hundred thousand Southerners fleeing a wall of roiling water eighty miles wide and up to thirty feet deep. The disaster was the result of more than two centuries of human failure to contain the ever-thrashing Mississippi River between its artificially heightened riverbanks.

The first levees erected to keep New Orleans dry were built in the early 1700s. By 1800, settlers had strung together a hodge-podge of differently sized and designed levees that stretched some hundred miles upriver from the city. The earthen berms proved no match for the chronically flooding river. Even after the government stepped in to build bigger and better-designed levees in the first half of the nineteenth century, the river continued to burst its manmade banks. It happened in 1844 and again in 1850 and again in 1858, and then again in 1862, 1867, and 1874.

The government's response to each flood was consistent. Up went the height and width of the sloped walls the engineers hoped would keep the river flowing south instead of naturally fanning across the ancient floodplain that was nature-designed to absorb and then slowly release the Mississippi's seasonal gluts.

Even with taller and stouter levees, the river chronically crested its artificial banks throughout the 1880s, 1890s, and the early 1900s.

Finally, when the deluge of 1927 hit, and the levees collapsed and unleashed the pent-up river's fury on the communities below as never before, the Army Corps surrendered. Or at least it retreated. It modified its levee-focused flood control policy and built the Bonnet Carré Spillway that functions something like a bathtub safety drain, scaled up to continental proportions. It was designed to shunt Mississippi River floodwaters into a relatively deserted swamp instead of drowning out the half-million downstream residents of New Orleans.

The 1.5-mile-wide Bonnet Carré Spillway stands today as it was built in 1931. Each of its 350 concrete bays is plugged with twenty vertical side-by-side wooden "needles" that are about eleven feet tall and a little thicker than a railroad tie. When floodwaters rise high enough against the structure, located about a quarter mile east of the normal river's edge, crews in bright orange vests use cranes to hoist the pins, one by one, to allow the swollen river to take a left and plunge into a levee-lined channel that flushes the overflows into Lake Pontchartrain, about six miles to the east.

Lake Pontchartrain isn't actually a lake, but an estuary in which smaller rivers east of the Mississippi mix with tidal waters of the Gulf before they are pulled out to sea. The outflow capacity from Lake Pontchartrain into the Gulf is so large that it can pass most anything the spillway can send it, so flooding in that area isn't a problem.

Completely opening the Bonnet Carré Spillway in this manner can take more than a week, but once all the pins are lifted, the spillway can pass enough water to fill three Olympic swimming pools every second. The structure isn't automated because the Army Corps never thought it would need to be opened more than once a decade or so. And that is one Army Corps calculation that proved spot on . . . at least for a while. During spillway's first fifty-two years, from 1931 to 1983, it only opened in seven different years.

But as the climate has changed in the past couple of decades, so has the operation of the spillway. Flood-driven high water forced the Army Corps to open it six times between 2008 and 2019. And in 2019 it was, for the first time, opened twice in one year. The first opening lasted 44 days between late February and early April as the continental United States wound up its wettest twelve months on record. The rains kept coming into late spring, so the guys in the orange vests lifted the spillway pins again in early May, and they stayed lifted until late July for an annual total of 122 days, a record duration and far beyond the 38-day average for all the previous years the spillway was opened.

The surge in spillway use in the past decade is not only the result of wetter weather but also a loss in precipitation-absorbing prairies, wetlands, and forests up north to make way for development, cornfields included.

This relationship between northern land use and southern flooding is lost on many Louisianians. Not Daryal Savoie. The sixty-two-year-old bayou native had some time to kill on a sunny late July day in 2019 while waiting for a pharmacy to fill a prescription, so he stopped by the Bonnet Carré to watch the Army Corps crew begin dropping the spillway pins back into place. He said he appreciated the structure's purpose and the work the guys in the orange vests were doing. "Nobody wants to see New Orleans flooded again," he told me. But, he added, that doesn't mean the spillway has come close to solving the problem of polluted floodwaters coming down from the north.

The retired truck driver spent twenty-two years crisscrossing the 1.2 million square miles of Mississippi River basin, so he knows its vastness in a way you can't grasp by looking at a map or even flying over it. He's rolled across North America's oceans of corn and soy and lumbered through storms that turned freshly planted fields into muddy morasses oozing freshly applied chemical fertilizer and manure.

"Everything up there," he said as he stood on the edge of the spillway, jabbing his thumb over his shoulder toward the Mississippi watershed that stretches north into Canada, "drains into this river, one way or the other."

Savoie owns a seventeen-foot Alumacraft fishing boat and

earlier that summer had taken it out on a flood-muddied Lake Pontchartrain, catching nothing but glimpses of dead fish bobbing to the surface because the spillway had turned the lake's naturally brackish water lethally fresh for its saltwater species.

The nutrient-loaded freshwater flows then made their way out of Lake Pontchartrain and into the Gulf, where they were suspected in the death of hundreds of bottlenose dolphins that washed ashore, their silvery skins pocked with brown lesions. The number of dead dolphins was so great that federal biologists declared it a "UME," which stands for an unusual mortality event.

What was also unusual was that the flows through the spillway were so large and lasted so long that they changed the water chemistry of the Gulf itself; in some coastal areas, the mid-summer salinity levels plunged to five parts per thousand, well below the thirty to thirty-five parts per thousand typical for ocean water.

These salinity readings weren't taken in the area where Lake Pontchartrain bleeds into the Gulf; one came from an Alabama research vessel sampling almost a hundred miles to the east, and about ten miles offshore. The scientist who took the reading called it "bonkers."

All this freshwater was trouble for more than just saltwater fish, oysters, dolphins, and sea turtles. It threatened humans because it triggered a phenomenon that scientists who have spent their careers studying the Gulf never expected: toxic freshwater algae blooms in the saltwater Gulf.

BY THE SUMMER OF 2017, the Gulf of Mexico, battered by 2005's Hurricane Katrina and ravaged five years later by the BP oil spill, was once again in a generous mood. The waters along Mississippi beaches suddenly burst with a miraculous abundance of sea life. Schools of fish roiled shin-deep waters, clusters of crabs shimmied up dock pilings. Shrimp swarmed in the bathtub-warm surf, and fevers of stingrays thrashed along the shore. The Mississippi Department of Marine Resources, a regulatory agency often criticized by fishermen for being too stingy in allocating the Gulf's bounty, spread the word that it was OK for the public to use nets, buckets, and even their bare hands to take whatever the sea was willing to give, catch limits be damned.

*Jubilee* is the word Mississippi biologists used to describe this nature-made buffet on their beaches. For most people the word connotes a 25th or 50th anniversary of some sort. But along Mississippi's Gulf Coast and Alabama's Mobile Bay, Jubilee has long described a cultural and culinary celebration of the Gulf delivering ashore a festival's worth of sea life ready to become seafood.

It is a joyous moment that rarely occurs in time (years can pass without one happening) and space—when one does occur, it is typically confined to narrow sections of beach along the Mississippi Sound, the barrier-island-protected ninety miles of coastline that begins east of New Orleans and stretches to Alabama's Mobile Bay.

The physics of the phenomenon that create a seafood Jubilee are not unlike the cause of the far larger dead zones menacing the Greater Gulf of Mexico. The difference is that a Jubilee is not driven by today's nutrient-fueled algae blooms or any other type of human pollution. Jubilees have been documented for more than a century; one local fisherman reported in 1960 that he had witnessed them for the past sixty years, and that his father had seen them all his life. Chuck Berry even sang about Jubilees in his 1957 classic *Rock and Roll Music*.

These spontaneous bonanzas are served up by a rare but natural mix of wind, water temperature, current, and tide. All conspire to suck pockets of lower oxygenated saltwater from near the ocean floor into the oxygen-rich surface waters along the coast. The upwelling of this water from the deep drives sea life toward the shoreline like a cattle stampede.

So long as the swarming fish, shrimp, and crabs are caught while flopping and flitting, they are perfectly safe to eat, according to the Mississippi Department of Marine Resources. "Currently our samples don't indicate that there are toxins present in the water, so the seafood likely is safe," the director of the department's Finfish Bureau said in a news release issued the day the 2017 Jubilee hit.

The official did offer one obvious caveat: "The seafood still should be handled, stored and cooked properly. Also, if any seafood is dead, and it looks like it's been dead for a while, it's best not to eat it."

Word of another Jubilee spread again exactly two years later in 2019 when tourists along the Mississippi coast started to post videos of spasming fish lips breaking the surface near the shore. This time, state regulators weren't about to sanction any scavenging, because these fish weren't trying to escape a natural phenomenon. They were fleeing waters that had, in almost an instant, turned as brilliantly green as antifreeze, and potentially just as poisonous.

"This is not a Jubilee," Emily Cotton, the chief of coastal monitoring for the Mississippi Department of Environmental Quality (DEQ) told anyone who would listen when the green water appeared in late June at the outset of Mississippi's peak tourist season. "It's a toxic algal bloom. Don't eat any of those fish. Just don't."

Two years earlier, Cotton had attended what she thought was a useless two-day seminar on how to identify and respond to blooms of a blue-green algae called Microcystis—the same type of toxic freshwater blue-green algae plaguing Lake Erie and so many other North American freshwater lakes, from New Hampshire to the Pacific Northwest.

Cotton's job, after all, is to monitor Mississippi's saltwater coast, and that means focusing on dangerous marine algae that cause red tides as well as fecal contaminants from sewage treatment plant overflows. "I'm never going to use this," she told herself as the seminar instructor went through the steps to identify a Microcystis outbreak, both on the water and under a microscope.

Then came the 2019 freshwater flood from the north that brought along a first-of-its-kind outbreak of Microcystis, which ecologists thought could not survive long along open coastal saltwater, let alone thrive to the point it posed a threat to human health.

Because of her training and the news that the Mississippi floodwaters had turned the Gulf Coast east of New Orleans into nearly freshwater, Cotton had a strong suspicion about what was happening when stretches of beach started to turn the telltale soupy green. She sent a batch of plastic jars containing the samples she took from the Gulf Coast to a state lab, and their test results did indeed indicate potentially hazardous levels of toxin-producing Microcystis.

On June 22, her department planted the first no-swimming signs on four of the state's Gulf Coast beaches. Cotton kept sampling in the following days, and she kept finding fresh Microcystis colonies. She closed five more beaches on June 24. She kept on sampling, and she kept on closing beaches. It wasn't long before she ran out of red and white no-swimming warning signs and had to get more printed.

By the end of the Fourth of July weekend, all twenty-one Gulf Coast beaches in Mississippi were posted as closed to swimming, just as air temperatures were about to hit one hundred degrees. The media spotlight was hotter. "Summer's the perfect time to hit the beach—unless you live in Mississippi," read one July 9 CNN story. The *New York Times*, CBS, NBC, and National Public

Radio ran similarly dire reports, all headlined with the tourism-killing words "Mississippi" and "Toxic Algae."

The beach closures were in full effect when I drove east on US 90 along the Mississippi coast in late July 2019. I found nothing but miles upon miles of deserted beach, as if the temperature under the hazy sky was forty degrees and not a searing ninety. Finally, about five miles west of Gulfport, Mississippi, I spotted a single person relaxing about twenty feet from the water.

Jill Wozniak had just arrived from her home in Lexington, Kentucky, after an all-day drive, a trip she and her husband take most every summer. She'd heard on the news that something was wrong with the water in Mississippi but it wasn't until she stopped to visit family on her way to the ocean that she learned *all* of Mississippi's beaches were closed due to toxic algae.

Once the couple got to Gulfport, Wozniak's husband had no interest spending an afternoon under a blazing sun with his feet stuck in dry sand. He opted for the hotel pool. Jill Wozniak was undaunted. She parked her car along East Beach Boulevard and walked straight past the NO SWIMMING sign and tried to make the best of the afternoon. She put on sunscreen. She pulled out her beach-read book. She uncapped her single-serve glass of pinot grigio.

Wozniak had been on the beach for about an hour by the time I walked up. "I would have canceled the trip had I known I couldn't go swimming," she told me. Then she confessed that she waded into the waves anyway. "I should know better," she said, explaining that she was a nurse practitioner. "But I didn't want to come all the way down here and *not* go in the ocean."

At that point there were no reports of illnesses related to swimming (of course, no one was swimming), and the scale of the algae blooms pocking the coast were tiny compared to the annual Lake Erie Microcystis blobs. Even so, the chief of the Mississippi DEQ ordered the no-swimming signs to stay up because lab tests consistently showed worrisome levels of the fast-growing blue-green algae that can produce colorless and odorless plumes of a class of toxins, called microcystin, that can break away from an algae bloom's telltale green slime.

State officials' caution to the public: just because you don't see a bloom on your stretch of beach doesn't necessarily mean the water is safe for contact because the toxin can be vodka-clear.

Wozniak found herself thinking just as cautiously after her swim. Her husband, she explained, had inherited a plot of land near that beach, and they had been talking about building a vacation home on it. "I'm thinking hard about that plan now," she said. "Do we really want to build something here when this kind of thing is going on?"

Wozniak was planning to return to Kentucky a couple of days later, leaving the Gulf's new algae troubles seven hundred miles behind. Not everyone, of course, had that luxury.

James Barney Foster has rented personal watercraft, sun umbrellas, and lounge chairs on the Mississippi coast since the 1980s. I was on my way to chat with one of his employees at one of his rental booths on Biloxi's East Central Beach when I got distracted by a historical marker. It turned out this beach is famous. It's where a thirty-one-year-old physician, Gilbert Mason, made international headlines in April 1960 after he was arrested and charged with disorderly conduct for jumping in the water. His crime: he was Black. The beach at that time was open only to whites. At least that's how the police saw things. Mason returned the next weekend with 125 volunteers for a peaceful "wade-in" to press the authorities to open the beach to everyone.

"Trained in non-violent passive resistance, they expected to be arrested," reads the plaque near Foster's jet boat rental stand. "Instead they were attacked by a white mob armed with pipes, chains, and lumber, while city police stood by without intervening." National outrage pressed the federal government to make things right in Mississippi; the US Department of Justice eventually stepped in and sued, a case that took nearly a decade to win.

Police returned to the same stretch of beach on July 3, 2019, and this time, because of the algae, they did not discriminate in ordering swimmers ashore.

"They come up on their four-wheelers yelling: 'Get out of the water! Get out of the water!'" Foster told me later in a phone interview. "It was crazy. It was just like right out of that movie *Jaws*."

The orders came at the worst possible time for Foster, and not just because it was the beginning of the long Fourth of July weekend. After a banner 2018 season, Foster had taken out a mammoth bank loan to purchase twenty-eight new high-end Yamaha WaveRunners he said had cost about a quarter-million dollars. The jet-powered boats were on their way to paying for themselves as Foster's rental business roared through the spring of 2019. Then came the algae, and then came the deputies. "This is worse for me than Hurricane Katrina. This is worse than the BP oil spill. I survived those," he said. "I don't know if I'll survive this."

Foster acknowledged that some beach closures along the Mississippi coast might have been justified, given the pictures he saw on the internet of algae blooms that were so eerily green that he didn't think it required signs or armed deputies to keep people out of the water. "I wouldn't jump in that. A fool wouldn't jump in that. But we've had none of that anywhere near us," he said in a rising voice. "I have never gotten sick from these waters and I'm 58 years old, a diabetic and I'm in and out of the water every day."

Foster was forced to chain-lock his fleet of giant-wheeled floating pedal bikes. He moved his WaveRunner rental operation to an inland bay a mile north of the beach that was so far off the tourist track that he only had about twenty customers over the entire Fourth of July holiday. In a normal year, he said, he might do fifty times that business. His beach chairs were still open for rentals, but on the day I stopped by his rental shack there were no takers. It was a similar story for businesses up and down the tourist-dependent Mississippi coast.

Mickey Bradley Jr. stood behind the counter in a gymnasium-sized gift shop called Sharkheads and did not try to gloss over what the beach closure had done to the Biloxi shop that sells T-shirts, swim shorts, seashells, key chains, fudge, and all sorts of other stuff people would buy on vacation. But almost no one was on vacation. "It's killing us," Bradley said. "You see it. It's barren out there. People are afraid."

Tourism officials spent the rest of the summer telling anyone who would listen that their beaches were not technically closed, that the swim ban was actually—legally—only an advisory, and that the sandy shoreline was still open for sunbathing, volleyball, and bonfires. But the damage had been done.

"We'll get over this, but people are angry right now," the public relations director for the tourism and marketing organization that represents Mississippi Gulf Coast counties told me after we both sat through a two-hour public hearing during which Gary Rikard, head of the Mississippi DEQ at the time, acknowledged the no-swimming signs were not likely to be lifted anytime soon that summer.

Foster was more than angry. He was packing. He needed cash—fast. He had a bank loan due the next week, and he didn't expect to get a break from his lenders in the uncertain environment southern Mississippi suddenly found itself in. His plan was to sell his new fleet of WaveRunners in Georgia.

"The banks don't want to lend you money, because the banks don't know what the Mississippi DEQ is going to do," Foster said.

DEQ's Rikard, an environmental attorney before his appointment to head the state environmental office, acknowledged at the public hearing that he could not guarantee similar closures wouldn't occur again in coming years. The problem, he explained, is that he can't tell the Army Corps how to operate the Bonnet Carré Spillway, and the pollution it is sending to the Mississippi coast is coming from farms in states far beyond his scope of authority.

IT TOOK THE federal government stepping in more than a half century ago to make Biloxi's beach safe for everyone, regardless of race, despite much resistance from white Mississippians. Now it appears it will take the federal government stepping in again to make the beach safe—from toxic algae blooms fueled by phosphorus pollution from the north.

It can't happen fast enough for Foster, who can't figure out why he has to pay the price for the mess northern farmers are making of his coastline.

"That's what they need to regulate—the people up there spreading the fertilizer," he told me. "Not us. Don't wait until it gets down here to start regulating."

The next spring the Army Corps was forced to open the Bonnet Carré Spillway once again.

It was a sign that until presidential candidates no longer have to profess their fealty to ethanol, or there are dramatic changes in the way agriculture is regulated, things are going to get worse for Mississippi.

And things are already worse along the Gulf Coast just to the east in Florida, where phosphorus-driven toxic algae outbreaks on both sides of the peninsula are no longer just affecting wildlife and holiday weekends—they're sending people to the hospital.

CHAPTER 8

# A Liquid Heart, Diseased

A LMOST SMACK in the middle of the southern Florida peninsula is a 730-square-mile inland sea known as Lake Okeechobee, second only to Lake Michigan as the largest natural freshwater lake lying entirely inside the continental United States. Lake O is nearly as round as its nickname, and more than thirty miles across. If you stand on its shoreline, you can't see to the other side. It feels vast as the ocean, but the reality is that today the lake functions more like a jumbo petri dish; Okeechobee is as shallow as a backyard swimming pool and just as warm. These characteristics make it a perfect incubator for phosphorus-fueled slicks of blue-green algae that don't dissipate as much as they metastasize down the manmade canals that send Okeechobee's toxic outflows toward seaside communities on both of Florida's coasts.

"Lake Okeechobee is like a giant cesspool, and then it goes east and west. And it's killing our estuaries," Jim Penix, who lives downstream from Okeechobee in the Atlantic coastal city of Port St. Lucie, told me during an extreme algae outbreak in 2018.

Okeechobee's woes are rooted in the vast agricultural lands straddling the tributaries that flow into it from the north. The basin's factory-scale dairies, sod and vegetable farms, sugarcane fields, and citrus groves all leach phosphorus into the ditches, rivulets, and rivers that drift toward Okeechobee. Mushrooming numbers of housing developments, commercial districts, and golf courses also send their phosphorus wastes into the lake, which, prior to construction of a giant earthen berm surrounding it, was as much a swamp as it was open water.

In its natural state, Lake Okeechobee's size, depth, and shape were in perpetual flux, swelling during the tropical storms and hurricanes that pummel the Florida peninsula every late summer, and shrinking back in the dry season. When the lake rose high enough, it would tilt over its southern shoreline and flush into a 50-mile-wide, 130-mile-long sheet of water drifting down the Florida peninsula and into the coastal waters at its southern tip that explorers named the "Ever Glades." These seasonal pulses from Okeechobee were the wellspring of Florida's famed River of Grass, and they were so regular they were almost rhythmic. They called Lake Okeechobee Florida's liquid heart.

Today, it is a heart that is severely diseased. The cure: cut phosphorus inputs to protect the lake's ecology, as well as the health of the roughly one hundred thousand people who live in the coastal cities straddling the canals that carry out to sea Okeechobee's toxic outflows. The state of Florida has a plan to do just this, but that's basically all it is—a plan, words on paper.

"The state has no hammer to make compliance happen. It's just a joke," John Cassani, head of a local environmental group in Fort Myers, a Gulf Coast city of about seventy thousand residents that is on the receiving end of Okeechobee's west-flowing toxic discharges, told me during an algae outbreak in summer 2018.

Things may be bad on the Mississippi coast and the western end of Lake Erie, but nothing compares to the tragedy of Lake Okeechobee.

The lake's story is one of relentless abuse inflicted by engineers trying to quell the lake's natural spillovers, of agricultural interests fixed on converting so much of the lake's wetlands into phosphorus-leaching croplands, and of politicians who have yet to muster the will to force the phosphorus polluters to change their water-poisoning ways.

It is a story that is resonant beyond Florida as we plunge deeper into the increasingly stormy and ever-warming twenty-first century, but its first chapter was written long ago, in the early twentieth century, on tombstones.

THERE ARE SEVERAL cereal box-sized grave markers clustered in a cemetery built on the slightest rise in the pond-flat farm country of central Florida. The aged markers just off Highway 78 are beginning to erode and tilt this way and that, haunted-house style. But you can still read the initials of the people buried beneath—E.M.B.; H.E.B.; W.J.B.; M.A.B.

The sameness of the markers and all those letter Bs are what make this forsaken plot forty miles east of Fort Myers particularly gloomy, even on a sunny July afternoon. That's because all the souls below belong to the same family. And they all took their last breaths together—on September 18, 1926, when Lake Okeechobee burst through a shoddy pile of dirt that was optimistically called a dike by the farmers who had been trying to scratch out a living on its dry side.

The nameless hurricane that pushed the lake over its man-made shoreline unleashed a muddy torrent fifteen feet deep upon the burgeoning farming city of Moore Haven and drowned hundreds, including E.M.B.—Eleanor Marie Blair—and her young children. All of them had sought shelter from the storm in a grocery store that collapsed in the swirling current.

Even as the wall of floodwater rolled for the southern tip of Florida, horror stories began to emerge from all the muck and splintered timber. One mother fashioned a raft from two automobile inner tubes and tried to ride out the swell with her daughters and infant son. The girls were peeled off by the frothing water just as they scrambled for a rooftop. The boy was swept away when the mother later tried to hand him to rescuers. Another mom lashed her toddler to a telephone pole at a height that she thought was sufficient to leave the child safely above the rising floodwaters. It wasn't.

Burials inside city limits were impossible because Moore Haven remained submerged under four feet of turbid flood-water even a week after the storm. This tendency for sheets of water to laze on the landscape in central Florida, in fact, is why settlers built the dike out of swamp muck and sand in the 1910s. The problem was that Lake Okeechobee's perfectly natural and predictable "floodwaters" were a menace to the farmers trying to coax crops of sugarcane, tomatoes, beans, potatoes, peppers, and eggplants from the soggy but remarkably fertile black soil south of the lake. The state of Florida spent some $15 million in the 1910s on a network of canals to drain down the lake so it would stop inundating all the newly cultivated farm fields. And, just in case the canals couldn't do the job, Floridians undertook a companion project to put Lake Okeechobee in its place— behind a chest-high dike.

But the dike didn't still the beating of Florida's liquid heart. It only quelled it until the pressure of suppressing more than a decade's worth of Okeechobee's freshwater pulses was unleashed in a matter of seconds and in a manner that instantly turned Moore Haven into what the press dubbed in the days after the 1926 disaster, "the city of the unburied dead."

Blame for relying on little more than a pile of dirt to check the immense forces that had shaped Florida over thousands upon thousands of years came swiftly and ferociously in the days after the flood, particularly from a local newspaper editor who argued for construction of a taller and thicker new dike that "would forever preclude such a tragedy which has recently in the space of a few hours changed Moore Haven from a peaceful farming community to a water-soaked tomb."

The dike was patched but not expanded in height. And almost two years to the day after the 1926 flood, another hurricane hit, and Lake Okeechobee once again heaved over the lowslung manmade berm at its southern end, this time about thirty miles southeast of Moore Haven.

Some two thousand people drowned in and around the town of Belle Glade, though the actual toll may have been much higher; many victims never surfaced and were presumed to have been inhumed in the mud as the lake's overflow made its run to the sea. Some drowning victims almost certainly were devoured by alligators. The remains of so many others were left baking in Belle Glade's streets and fields, their stiff corpses rotting in the early autumn heat in a scene so horrific that the Miami *News* described it as "too gruesome for a newspaper story."

Many white victims eventually received proper burials, but many of the bodies of Black farmworkers were stacked in piles, doused with fuel, and set aflame. Their charred remains were piled into mass graves, one of which today reportedly holds the remains of some 1,600 souls just east of Lake Okeechobee, on a patch of land not much bigger than the footprint of a motel room.

This second flood was so catastrophic that it caught the attention of president-elect Herbert Hoover, who arrived soon after in a twenty-car caravan to tour the devastation. The Stanford-educated engineer-turned-politician was reportedly left in tears as he promised the survivors federal help was on the way. Within a decade the US Army Corps of Engineers had completed a massive upgrade of Okeechobee's flood-control structures. It included an expanded canal system to channel Okeechobee overflows to both of Florida's coasts, along with a system of fatter, higher dikes. A side benefit of all the ditch-digging and dirt-piling was that it created a navigation corridor running from the Atlantic coast city of Stuart across the Florida peninsula—including across Lake Okeechobee—and over to Fort Myers on the Gulf Coast.

Predictably, the new safeguards proved inadequate when another hurricane hit in 1947. The dikes held, barely, but the canal system proved too feeble to carry the floodwaters out to sea. The result was the worst flooding on record at the time to hit south Florida, in terms of acres submerged.

Predictably, the devastating inundation prompted calls for a new, grander dike system, one that would silence Okeechobee's wild heart once and for all.

Predictably, more earth-moving equipment rolled toward the Everglades once again.

"This monster had to be controlled by bigger levees and by bigger canals that would give it bigger outlets to the sea," the US Army Corps of Engineers proclaimed in a 1950s documentary about the agency's campaign to "fix" the lake and drain some fifteen thousand square miles of Florida swamp. The project included expansion of the canals running east and west from the lake as well as construction of a three-story-high dike that ringed the lake's 143-mile circumference.

Finally, the Army Corps crowed, humans really did have Mother Nature in shackles.

"Water that once ran wild; water that ruined the rich terrain; water that took lives and land and put disaster in headlines and death upon the soil . . . now it just waits there—calm, peaceful, ready to do the bidding of man," brayed the narrator of the film the Army Corps titled *Waters of Destiny*. "Central and southern

Florida is no longer nature's fool, the stooge for the practical jokes of the elements!"

Converting the northern Everglades into a sea of sugarcane, tackling Okeechobee's deadly flood problem, and carving a bigger navigation channel across the Florida peninsula might have looked like a win-win-win on the drafting boards of mid-twentieth-century engineers, but on the ground in central Florida today it is shaping up as an unbridled disaster.

AS REAL ESTATE DEVELOPMENT and farm operations expanded into Florida's once-untamed interior in the second half of the twentieth century, Lake Okeechobee began receiving massive amounts of phosphorus—concentrations of the nutrient in the lake roughly doubled from the 1970s to the early 2000s.

Today the annual load of phosphorus tumbling into Okeechobee from its tributaries can be as high as 2.3 million pounds—about ten times what biologists estimate Lake Okeechobee can safely absorb before it begins to cough up dangerous amounts of toxic algae. Like in other watersheds across the country, much of the phosphorus load flowing into the lake can be traced straight back to dairies and crop-growing farmers inside the watershed.

But open-range beef cattle ranches are a different story, and Florida is, surprisingly, big cowboy country. Some of the largest cattle ranches in the United States are in Florida, including one ranch near Orlando that sprawls across three hundred thousand acres—twenty times the size of Manhattan. Yet the state's vast ranchlands aren't the agricultural pollution menace many assume them to be.

Take rancher Wes Williamson, who has lived on his family's ten-thousand-acre ranch north of Okeechobee almost all his life. On a good day, the sixty-something roams his pastures in a

Ford F-Series pickup, which is old enough to still have a Romney-Ryan bumper sticker from the 2012 election. On better days, he goes deeper into the bush on his Polaris four-wheeler. On the best days, he's riding his quarter horse named Blue, rounding up calves. I met Williamson on one of his best days.

He'd been on his horse since 6 a.m., corralling two hundred calves to ship to out-of-state feedlots. Williamson explained that he uses some phosphorus fertilizer to boost grass production on his range. He also gives his herd a nutritional bump with feeds derived from things like orange peels, dried grain from ethanol plants, and cotton seed, all of which contain phosphorus. But he also exports his share of phosphorus out of the watershed—pointing out the thousands of pounds of livestock he shipped out earlier that day.

Williamson said he also does what he can to keep manure and other nutrients out of his creeks that drift toward Lake Okeechobee. That includes converting some 2,500 acres of pastureland into non-grazed wetlands that are intended to capture phosphorus. He figures the best thing he can do to help Lake Okeechobee is to keep his land in pasture, and keep the developers away.

"We're not just cattle ranchers," he told me, "we're grass farmers."

Thirteen million people live within a two-hour drive of his ranch, and housing developments are inexorably creeping inland.

In 1960, Florida had fewer than five million residents. Today it has some twenty-two million. Nearly one thousand people move to Florida every day, with the state expecting to add as many as five million residents in the next decade. Many of them will move inland to the Okeechobee watershed, where their collective wastes will likely only exacerbate the lake's phosphorus woes.

"When a rancher sells to a developer," says Williamson, "the developer plants the last crop. And that's houses."

An equally daunting problem is that the dike the Army Corps built to contain Lake Okeechobee is, literally, a pile of . . . stuff.

The steeply pitched grass-covered mound of crushed seashells, dirt, sand, and rocks that holds Lake Okeechobee in place today is known as the Herbert Hoover Dike. It was completed a few decades after Nevada's Hoover Dam opened in 1936. Florida got the poor man's Hoover, one constructed with neither the Deco flair possessed by its concrete cousin nor with the appropriate engineering to withstand what Florida weather can throw at it.

The Hoover Dike, in fact, wasn't designed as much as it was just piled high. More worrisome, the dike (which the Army Corps also describes as a "levee") has for decades been used for something it was never designed to be—a makeshift dam to hold back Okeechobee's water in dry years to provide irrigation water for the sugarcane fields that dominate the landscape to its south. There is a critical distinction between a dam and a levee. Levees, like a stack of sandbags, are designed to hold back floodwaters on an emergency basis. A dam is a meticulously designed structure of concrete or highly compressed earth built to hold water back constantly.

Expecting a levee to function as a dam is like asking a paper cup to do the job of a coffee mug. That includes expecting it to withstand repeated cycles through a dishwasher as if it were made of ceramic, not pulp.

The Army Corps is shockingly blunt about the vulnerabilities of the thirty-foot-high Hoover Dike, which today is the only thing separating the lake from the living rooms of the tens of thousands of Floridians who now live downstream.

"The problem with the Herbert Hoover Dike when it comes to high water levels can be summed up in two words," the Army Corps acknowledged in one report, "It leaks."

A team of risk assessment experts for Lloyds of London toured the Hoover Dike after Hurricane Katrina to assess its vulnerability to storm-driven floods. They walked away from the "naturally porous" structure plainly concerned, and not just about the dike being overtopped during a big storm. "Eventually," the Lloyds inspectors predicted, "either the constant water pressure on the dike wall causes its slopes to fail, or it eventually collapses from the net effect of particle removal of one grain at a time."

The Army Corps is in the middle of a $1.7 billion project to fortify the dike with a concrete and steel spine in its particularly vulnerable sections. The agency is also rebuilding the erosion-prone structures that control water releases from the lake. The project isn't expected to be completed until the mid-2020s.

In the meantime, the Army Corps tries to keep pressure off the dike by doing what it must to try to keep the lake's surface elevation from surpassing 15.5 feet above sea level, well below the top of the thirty-foot-high structure. The Army Corps calculates that the dike enters the danger zone for failure if the water climbs to more than 18.5 feet above sea level. If the level reaches 21 feet, still well below the dike's top, the agency says a collapse is "likely" and could come with little warning, or no warning at all. The odds of the water reaching that height in any given year: a not-so-comforting one in a hundred.

"With 40,000 people living in the communities protected by Herbert Hoover Dike," the Army Corps acknowledges, "the potential for human suffering and loss would be significant."

Like Civil War troops stacking pyramids of cannon balls in advance of battle, Army Corps workers stockpile boulders and different-sized rocks placed in strategic areas atop the dike for use as emergency plugs when leaks spring. And because the lake can rise so suddenly, particularly during hurricane season, the Army Corps has a practice of opening a set of flood gates on the dike in mid-summer to send lake water to the Atlantic and Gulf coasts, even if the lake level isn't even close to climbing into the danger zone.

The rationale for these non-flood releases is that the Army Corps needs plenty of space behind the dike in advance of the peak hurricane season that comes in late summer.

It's a game of roulette the Army Corps may eventually lose. The basic problem is that the sprawling landmass that drains into Lake Okeechobee covers roughly 4,400 square miles, more than six times larger than the surface of the lake itself. Drainage from a landmass that size means the lake's inflow can be far larger than the capacity of the manmade canals built to carry away rising waters; when a string of big storms hit the lake can rise as much as four feet in a month. So even just a few wet weeks can take the lake from safely low levels to the danger zone.

Had Lake Okeechobee and the northern Everglades been left in their natural state, few people would be living on—and farming—the lands in the lake's flood path. Much of the algae-fueling phosphorus now trapped behind the dike would instead spill over the lake's natural southern shoreline and be absorbed, kidney-like, by the Everglades. And any freshwater algae blooms that did survive a trip from Okeechobee through the glades would flush out into the sparsely populated southern tip of Florida, where waves would break up the blooms and saltwater would eventually kill them.

But Lake Okeechobee, of course, was not left in its natural state, and now the Army Corps is trying to engineer its way out of its original "fix."

One component of the ongoing multibillion dollar Everglades restoration project, jointly funded by the federal government and the state of Florida, is a plan to dig a mammoth, $3 billion reservoir south of Okeechobee that would capture the lake's phosphorus-rich and algae-contaminated overflows and then slowly, safely release that water into the Everglades. The project would reduce or eliminate today's practice of shunting Okeechobee's contaminated flows down canals to Florida's heavily populated coasts. But it has not been funded, and even if it is, it likely won't be completed for a decade or more.

So contaminated water releases to Florida's coasts through the Okeechobee canals will continue for the foreseeable future, no matter how polluted the lake water might be, and it can be extremely polluted.

At one point in mid-summer 2018, 90 percent of Lake Okeechobee's 730 square miles were coated in a gooey layer of blue-green algae dense enough for armadillos to toddle across. The water level of the lake at the time was nowhere close to over-topping the dike, but the Army Corps nevertheless opened the gates and flushed those fouled waters into the canals that connect to the Gulf-bound Caloosahatchee River and the Atlantic-bound St. Lucie River.

Florida's governor accused the Army Corps of mismanagement for discharging Okeechobee's toxic water in a manner that not only threatened aquatic life but also the people who live downstream along the canals, river channels, and coastlines hit with muck. The Army Corps insisted it needed to keep the lake low in case a severe storm season hit, so it did not have a choice in the matter.

It didn't.


I TRAVELED TO FLORIDA in summer 2018 at the peak of the contaminated releases from Lake Okeechobee and just days after then-governor Rick Scott declared an algae-fueled state of emergency. My first stop was Fort Myers to meet with John Cassani, head of the local chapter of the national Riverkeeper conservation organization.

I was hoping to go out on a boat with him on the Caloosahatchee River that receives canal flows from Okeechobee and then discharges them into the Gulf. I wanted to witness the blooms for myself, and I was encouraged when Cassani arrived wearing boat shoes and a long-sleeved sun-proof shirt. He looked ready to sail. Then he told me he wasn't about to go out on the river—not with me, not with anyone. The closest he was willing to get to the water, in fact, was a Cracker Barrel along Interstate 75 in Fort Myers.

When the summer season's first swirls of Okeechobee's toxic algae had drifted down the Caloosahatchee a couple of weeks earlier, Cassani was eager to take reporters out on the water to explain the origin of the not-so-natural disaster that had happened three out of the past four years. But soon enough he was left with burning lungs, itchy eyes, and an un-kickable dry cough from the fumes floating off the rotting algae. He described the stench as "kind of like a mixture of baby diapers and moldy bread." The problem wasn't so much the stink but how his body had started to react to it. "I just start gagging," he said, "and I want to vomit."

Blue-green algae blooms are particularly dangerous when they begin to rot because the toxins they carry are released as the individual cell walls die. The fumes got so pungent with the 2018 outbreak that one of Cassani's colleagues was forced to move out of his home on a canal branching off the river. "He just couldn't stay there with his family," Cassani told me. "It wasn't safe."

Cassani explained he had come to expect late-summer algae troubles from Okeechobee, but 2018 was different. The first poisonous plume arrived in Fort Myers less than two weeks into summer. This left Cassani believing that his hometown is headed for a new normal, one in which its coastal waters could be off-limits for much of the summers.

"That's what has everybody freaking out. It's just too early," he told me as he glumly picked at a fruit cup in the Cracker Barrel dining room, his back to a wall decorated with old-style fishing gear and black-and-white fishing expedition photos.

An antique bedpan hung from the ceiling near a cashier station at the restaurant—apparently Cracker Barrel's crack at evoking the ambiance of an Old Tyme country store. The kind of store you might actually have found in Fort Myers in the first half of the twentieth century, when the Caloosahatchee River ran clear to the coast from Florida's interior, and the city and its surrounding Lee County had a population of less than twenty-five thousand. Today the population is pushing about three-quarters of a million, and demographers predict its number of residents could nearly double in a generation.

Cassani smiled about the bedpan. Fecal-inspired décor may not be entirely appropriate for a restaurant, but he found it apt for Florida as a whole, because the state is awash in a steady stream of waste that it just can't handle. Even so, Cassani noted the state had recently passed a law giving environmental regulators a twenty-year extension to bring the phosphorus loading of Lake Okeechobee down to safer levels.

Florida, he explained, is actually going backward with its water quality protections at the same time its residents are increasingly choking on the fumes coming off the water.

"Things are thoroughly screwed up. Thoroughly," Cassani said.

And blue-green algae isn't the Gulf Coast's only algae problem. There is concern phosphorus might also be a factor in the Gulf Coast's menacing red tides, though those saltwater algae blooms have been documented for centuries, and scientists note outbreaks of the algae that causes red tides can start up to forty miles from shore—largely beyond the reach of coastal nutrient pollutants. Still, fertilizer pollution, particularly nitrogen, may exacerbate the rusty-colored plumes as they drift shoreward.

I left Cassani at the Cracker Barrel and drove to the mouth of the yawning Caloosahatchee River near where it drifts into the Gulf of Mexico. The gooey algae mats that had made national headlines had broken up a day or two earlier, but marble-sized algae globules still drifted just below the surface of a river carrying so much churned-up sediment that it was almost black as coffee. There wasn't a boat on the water.

I headed inland for Lake Okeechobee, where I met Paul Gray, a biologist with Florida Audubon, in the parking lot of a Best Western about a half mile north of the lake. The first thing he wanted me to know was that we were standing on the original lakebed. Today, because of the dike that raised the lake's water level but shrank its footprint, the former lake bottom and wetlands on the northern edge of Lake Okeechobee now sprout the kind of big box businesses you expect in a booming state—a Home Depot, a Walmart, a Publix grocery store, as well as an airport.

It was in this place, just a little east of what is now a McDonald's parking lot, that on Christmas Day 1837, Colonel Zachary Taylor marched some one thousand troops down to the northern shore of the lake. The future US president then ordered 132 Missouri Volunteers to make a run at the hundreds of Seminoles encamped on a patch of high ground between Taylor's troops and the open waters of Okeechobee. Taylor had the soldier numbers and the literal horsepower to surround his enemy in what turned out to be a pivotal battle of the Second Seminole War, but he instead ordered his troops to attack on foot and head-on—through waist-deep mud. The Seminoles took advantage of their elevated position and killed almost every officer before they disappeared into the tall grass.

Both sides would later claim victory, but Taylor's men suffered far more casualties. One of the reasons for the lopsided outcome was the Seminoles had long ago learned to live with the swamp instead of fighting it. It is a lesson the US Army—more specifically, the US Army Corps—has, manifestly, yet to take in.

Gray wanted me to see for myself how an irrigation and water management system that was designed seventy years ago to open hundreds of thousands of swampy acres to agriculture and to protect vast swaths of southern Florida from inundation is now failing miserably. "It's like we're driving a 1940s-era car," he said, wheeling his Prius past the Seminole battlefield that is now a state park, "and it's going to cost billions of dollars to get it fixed."

Billions of dollars Florida politicians so far are not willing to spend.

Our first stop was at the mouth of Lake Okeechobee's eastern canal that sends lake water through its flood control gates and into the St. Lucie River that runs through the city of Stuart, about thirty miles to the east. In the days before I arrived, the Army Corps had been spilling almost two billion gallons of blue-green algae-contaminated water per day through those flood gates, and another three billion gallons per day down the Fort Myers-bound canal on the other side of the lake.

"IF I WERE the Corps I'd do the same thing," Gray said as we stood atop the dike and gazed down on "water" as green and as pasty as guacamole. "You can't afford to be an environmentalist when the dike might break."

A slight breeze kicked up the algae fumes, the first whiffs of which smelled like freshly cut grass. Then a burn in my chest set in, and I began to cough as we walked along the lakeshore. I wondered if it was all in my head, but some of it had to be in my lungs; the cough persisted for a couple of days.

Gray, a Missouri native with a doctorate in conservation biology, never expected to make Florida his permanent home when he arrived as a graduate student in the 1980s, but the place grabbed him. "I grew up on a prairie and I never even knew it, because by the time I was born it was just cornfields," he said, adding that if he wants to catch the faintest sense of what that long-gone prairie looked like, he has only paintings to look at.

Or he can take a trip to Missouri's Prairie State Park. It would take a motivated cowboy about ten minutes to cross that patch of grass on horseback.

What he was trying to tell me is that Missouri had been tamed in a way that central Florida still has not.

Floridians might have remade nearly five hundred thousand acres of its northern Everglades into a sea of sugarcane and choked the natural seasonal overflows from Okeechobee along the way, but the central peninsula remains one of the most wildlife-filled—and wildest—places in the lower forty-eight.

Gray pointed to the dizzying number of species that still call Lake Okeechobee home, despite the summer algae outbreaks. There are thousands of wading birds—snowy egrets, great blue herons, and wood storks. The skies above the lake swarm with ducks—diving ducks, dabbling ducks, wood ducks, and mottled ducks. In that same airspace darts the predacious orange-beaked snail kite, and below the waterline dwell the apple snails that the endangered raptor relies upon. The lake's marshes teem with frogs, turtles, snakes, lizards, and alligators. Otters prowl waters still thick with shad, bluegill and black crappie, eels, sunfish and chubsuckers.

More than this, there is the awesome weather that still has its way with Florida, and still has the Army Corps scrambling to control what has so far proven to be uncontrollable.

While Gray was distressed that the Army Corps was sending another toxic algae blob from Florida's outback to its heavily populated coasts, he said as long as water from Lake Okeechobee must be channeled to the east and west coasts to prevent a dike collapse, and as long as the lake is continually oversaturated with phosphorus, summertime toxic algae is to be expected on Florida's coasts.

And even if all phosphorus discharges were halted tomorrow, he explained, there is still so much of the stuff piled up on the lake bottom and baked into the landscape draining into the lake that it could take decades for Okeechobee's waters to recover. Meanwhile the nutrient overdosing goes on.

"We keep adding more phosphorus and we keep expecting water quality to get better. It's the kind of thing a kindergartener wouldn't believe, but everybody else seems to," said Gray. "Every day we expect things to get better, and every day they are actually getting worse."

In the sky above us, a drone buzzed the Okeechobee shoreline, evidently capturing images of the slug of blue-green algae headed for Stuart. With its pilot nowhere in sight, Gray joked that whoever was flying it might have been doing so from the comfort of their living room in Stuart. This was not so far-fetched because residents of the little city downstream from Okeechobee and about forty miles north of West Palm Beach are becoming algae bloom-consumed.

LESS THAN AN HOUR drive to the east, a monthly meeting was taking place of the Rivers Coalition, a swelling group of Atlantic coast residents demanding an end to the toxic canal flows. The gathering in Stuart City Hall started with the Pledge of Allegiance, and then the attendees introduced themselves and their respective affiliations, and each introduction was acknowledged with a smattering of polite claps. A representative from Republican senator Marco Rubio's office showed up, as did a representative for Republican congressman Brian Mast. There was a lady from the League of Women Voters and a high school teacher running for state office. Some attendees represented homeowners' associations. Two people came to speak up for their yacht clubs. A fishing club, a sailing club, and a rowing club were all represented, as was a member of the Martin County Farm Bureau. One man characterized himself as simply a "very pissed-off resident." He probably got the most applause.

The focus of the Rivers Coalition's fight is to force the Army Corps to allow more Okeechobee water to run its natural course south toward the Everglades instead of shunting it into the coastbound canals. That can't happen in a major way until sugarcane fields are taken out of production and the planned reservoir is built to hold Okeechobee's overflows, a project that has been stalled for more than a decade.

So Lake Okeechobee continues to suffer in a manner that has long consumed environmentalists but remained mostly out of sight and out of mind for the millions of homeowners and visitors who rarely venture more than five miles inland from the beach. That's changing now with Floridians awakening to the idea that the health of the state's liquid heart is tied to the health of its toney oceanfront neighborhoods, to the health of its coastal fishery, and to the health of themselves and their children.

One meeting attendee was Blair Wickstrom, the publisher of the magazine *Florida Sportsman.* The algae forced him to evacuate his six-thousand-square-foot office on one of Stuart's most polluted canals, where a mound of rotting blue-green algae, also called cyanobacteria, had piled so high that it looked like you could roll a bowling ball across it. The crusty gunk had begun to turn turquoise—the telltale sign that the cyanobacteria cells are dying and unleashing their toxins into the water and air.

Wickstrom left a sign on his office door: Closed—Toxic Algae Fumes. He showed up at the Rivers Coalition meeting with eyes red as a pallbearer's and a stomachache that he said had him eating nothing but Tums antacid for three days straight. He likened the poisonous algae blooms to Cleveland's Cuyahoga River fire in 1969, which is credited for sparking public outrage that helped lead to the passage of the Clean Water Act. "It's not like I'm glad we have toxic water here," he told me, "but it is giving us attention that we wouldn't have if it was just killing birds, fish, crabs, and manatees."

A sore throat is not unusual for people exposed to the fumes from the blooms of cyanobacteria plaguing Stuart known as Microcystis, the same stuff closing Mississippi beaches. Other ailments triggered by short-term exposure include vomiting, dry cough, pneumonia, and abdominal pain.

More ominously, some of the health impacts tied to chronic exposure might take decades to manifest.

Years of sustained exposure to the type of toxins produced by Microcystis, called microcystin, has been linked to nonalcoholic liver disease and even liver cancer, but Dartmouth neurologist Elijah Stommel has bigger worries. He specializes in treating patients with amyotrophic lateral sclerosis, also known as ALS, or Lou Gehrig's disease. The terrifying affliction attacks the motor neurons in the brain and spinal cord that control voluntary muscles. Death is almost always the end result, and is usually caused by respiratory failure, typically just a few years after a patient heads to the doctor with vague complaints of things like shortness of breath, slurred speech, swallowing problems, muscle cramps, and twitches or numbness in the arms and legs.

It's an uncommon disease, but not uncommon enough; even working in lightly populated New Hampshire, Stommel loses a patient to the disease every couple of weeks, so many that in 2008 he decided to map where they lived.

One of the most unnerving aspects of ALS is that it seems to strike so randomly. Researchers estimate that 5 to 10 percent of ALS patients can track their disease back to their genes because it has been demonstrated that occasionally it can run in families. But for most of the afflicted, there is no apparent genetic link, and researchers believe the disease may be caused, at least in part, by some sort of environmental trigger—a stealthy toxin (or toxins) that has yet to be conclusively identified. Stommel is hunting for that toxin, and that is why several years back he had a couple of his students map the addresses of his patients on a Google Earth program to hunt for potential environmental factors.

"I wanted to see where they lived and what they might be exposed to, and I noticed many lived around Mascoma Lake," Stommel said. The lake in western New Hampshire, prone to blooms of Microcystis, is about four miles long and a half mile wide and is mostly located in the little town of Enfield, New Hampshire, population less than five thousand.

ALS typically strikes about two people per one hundred thousand, yet Stommel found he had nine patients who lived in that town alone in 2008. That spike in ALS—Stommel said the rate appears to be about twenty-five times greater than would be expected for a town that size—could plainly be a statistical quirk. Still, the numbers got Stommel thinking about the lake's summertime cyanobacteria blooms, and that got him thinking about . . . Guam.

The US island territory in the western Pacific has been intensely studied since the 1950s after researchers identified a remarkably high rate of an ALS-like disease in a group of native Guamanians known as Chamorros. The islanders suffered from the disease at a rate up to one hundred times that of what would be expected. Scientists, reckoning that studying the outbreak might lead them to finding an environmental trigger for ALS, descended on Guam and quickly focused on the Chamorros' diet. It largely consisted of a type of tortilla made by crushing the plum-sized seeds of the island's palm-like cycad plants.

Researchers analyzed cycad seeds and found they contained substantial amounts of an amino acid called BMAA, which lab experiments showed could ravage nerve cells in a petri dish. But after experiments in which rats were fed the BMAA stripped from the cycad seeds, scientists calculated it would require a human to eat thousands of pounds of cycad seed tortillas to reach levels high enough to show any signs of neural damage.

The BMAA-brain disease hypothesis faded in the 1970s but then flickered back into public consciousness in the early 2000s thanks to a Harvard-trained ethnobotanist who proposed a different pathway for BMAA to have ravaged the Guamanians' brains. After spending extensive time with the Chamorros, he learned that they had long feasted on giant "fox bats" until their numbers crashed due to overhunting in the decades following World War II. Those bats, it turned out, feasted on cycad seeds. The botanist theorized that, over time, BMAA accumulated in bat brains in far greater concentrations than could be found in the seeds themselves. And when the bats were cooked whole—head and all—in coconut milk and served as a stew, he theorized that might provide a strong-enough dose of BMAA to ravage, or play a role in ravaging, the brains of those who regularly feasted upon them.

Furthermore, he dug into the roots of the cycad trees and found they were laced with cyanobacteria that happened to be rich in BMAA.

The article he published in the journal *Acta Neurologica Scandinavica*, coauthored by the famed neurologist and author Oliver Sacks, revealed the potential mechanism for the toxin to make its way into the cycad seeds, and then into the bats, and then into the humans who ate them. The idea isn't that someone who eats a bowl of bad bat stew will come down with ALS the next day, or even the next year. It is a slow process that would take decades to manifest, which also might explain why the Guam disease rate has faded in recent decades, right along with the bat population.

The theory was received with tremendous skepticism from many in the medical community, but in the nearly two decades since that journal article was released, the bat BMAA hypothesis has gained some—but far from universal—acceptance.

One believer is Larry Brand, a professor of marine biology and ecology at Florida's University of Miami. "When I first read about the bats I thought—that's kind of a weird, isolated and unfortunate incident in Guam," Brand told me. Then he talked to a neurologist at his university who directed its "brain bank"— one of six National Institutes of Health biorepositories set up to collect donor brains for research into neurological diseases and other ailments. It turned out that she had gone looking for BMAA in the brains of some donors who had died of ALS and Alzheimer's, and she found it.

"At that point," said Brand, "I'm thinking this is not just a problem about bats in Guam." The Guam BMAA, after all, was believed to have come from cyanobacteria lacing the roots of cycads. But Florida also has loads of cyanobacteria due its chronic algae blooms. So Brand went looking in 2010 to see if the increasing blooms were also releasing BMAA, and he found it literally swimming in the waters around South Florida.

"I discovered high concentrations of BMAA in the food chain," he said. "In shrimp, crabs, and bottom dwelling fish like puffers. In some of these shrimp and crabs you can find BMAA in just as high a concentration—and sometimes twice as high— as the bats in Guam."

A 2020 University of Miami analysis of dead beached dolphins showed their brains had BMAA levels similar to those found in the brains of donors to the Miami brain bank who had died of ALS. The dolphins weren't killed for the experiment; they were collected dead on the beach. And of the thirteen tested, twelve of them had BMAA in the brain. The one dolphin that didn't had been killed by the propeller of a boat motor.

So what does this all mean for humans who might be exposed to the BMAA through seafood, drinking water, or even breathing it in if wind and waves kick it into the air? Nobody is predicting that people on the Florida peninsula or anywhere else are suddenly going to start suffering ALS, but some scientists are beginning to wonder if a surge arrives in the coming years or decades, particularly as the algae blooms grow in intensity.

Still, plenty of scientists have written off Stommel's New England lake work as a statistical hiccup, a textbook example of confusing correlation with causation. Stommel is clear that he's not trying to convince anyone they are going to die if they live near a lake prone to blue-green algae outbreaks. But he says it could be a contributing factor in a highly complicated disease equation involving one or more environmental toxins, genetics, lifestyle, and even simple bad luck.

"Not everyone who is living on a lake (with toxic algae blooms) is going to come down with ALS, and not everyone who eats fruit bats in Guam is going to come down with ALS," he says. "But if you have the right genetic predisposition, you are more likely to. It's like smoking and lung cancer. Not everyone who smokes gets lung cancer."

Todd Miller, a University of Wisconsin-Milwaukee researcher who focuses on the toxicity of cyanobacteria blooms, picks his words diplomatically when he calls the idea that algae blooms could somehow be a trigger for ALS "highly controversial."

"I don't doubt they've detected BMAA in dolphins and other biota," he says. "I do think there is still a question about how much of it is required to cause neurodegenerative effects."

Yet even if the science on the cyanobacteria-ALS connection remains hotly controversial, there is little question the ongoing phosphorus overdosing of Lake Okeechobee is going to keep the toxic algae blooms coming, right along with health complaints.

During my trip to Stuart I visited Tom Cubr, a salesman at a marina and boat dealership whose inventory listed everything from $100,000 recreational fishing boats to a $4 million yacht. Cubr said the raging algae bloom was dreadful for his business, but he worried about more than sagging sales. He fretted for his own health, explaining that he tried to ignore a sore throat that had hit a few days earlier, before he admitted the algae fumes coming off the canal outside his office were the likely problem. "I thought it was my imagination," he said. "It wasn't my imagination."

He explained that a similar cyanobacteria outbreak in 2016 didn't just drive away boat buyers. It also jeopardized the reputation of Stuart—"The Sailfish Capital of the World"—as a fishing and yachting destination. That year, he said, the St. Lucie River estuary seemed to lose all its fish for several weeks until winds, waves, and tides finally pulled the toxic slicks out to sea.

"Two months later the porpoises were back, the seagulls were back, and it's because the fish came back," he said. "Mother Nature, she did a great job of naturally cleaning it up. But how many times will she do that?"

Cubr pointed out that the marina workers cleaning boats outside his office window had started to wear the same type of respirators fiberglass workers use. They also wore protective glasses and rubber gloves. He shook his head as he told me he had just heard that morning that the Army Corps was about to begin increasing contaminated Lake Okeechobee flows to Stuart.

"I've never thought of myself as an environmentalist," he told me, "but you don't have to be an environmentalist to see that this thing is wrong and it needs to be solved."

*Part III*

# THE FUTURE
## OF
## PHOSPHORUS

CHAPTER 9

# *Waste Not*

A N EXQUISITELY BALANCED phosphorus exchange
existed for billions of years before humans corrupted
the element's flow through the environment. The initial phosphorus atoms that trickled from Earth's solidified magma into
the living world became building blocks in the first single-celled
organisms as life took hold in the ocean. As more and more
phosphorus escaped those slabs of igneous rock, more life—and
more complex forms of it—surged around the globe, first in
the oceans and eventually on land where, as in the sea, eroding
rocks leached traces of the precious element necessary for every
living thing.

The phosphorus atoms driving all this life bounded between
land and sea. Some phosphorus on the landscape washed away
with soils or their dead host organisms into rivers, lakes, and
oceans, where it was free to cycle around and around the aquatic
food web.

Waterborne phosphorus sometimes flowed in the opposite direction as algae washed ashore and its phosphorus was

absorbed by terrestrial plant life. Or phosphorus could make its way inland in the form of massive fish migrations up coastal tributaries, where spawning fish became an easy target for all manner of terrestrial scavengers and predators.

Whether in water, on land, or toggling between the two, the phosphorus atoms let loose in the world entered a timeless ticktock between the living and the lifeless, a dynamic that humans have intuited since antiquity; it is the literal manifestation of the Biblical notion of ashes to ashes. Or, as Joni Mitchell sang it: We are stardust. (There is evidence, in fact, that some of the Earth's phosphorus may have been delivered by meteors.)

Some phosphorus ended up in organisms that expired and drifted down into the relatively lifeless universe of the deep ocean, but those losses were offset by the steady stream of phosphorus released from weathering rock.

It wasn't, of course, a perfect equilibrium—one atom lost to the ocean floor, one atom set loose from igneous rock weathering. But a NASA-funded study conducted several years ago suggests how elegantly tuned aspects of Earth's natural phosphorus flow likely were before humans hijacked it. Researchers working with satellite data analyzed the tonnage of phosphorus in dust clouds billowing off the Sahara Desert and drifting westward across the Atlantic Ocean to the Amazon jungle, a jet stream–forged bond between one of the planet's most desiccated places and one of its most verdant—and surprisingly phosphorus-hungry.

"Nutrients—the same ones found in commercial fertilizers—are in short supply in Amazonian soils. Instead they are locked up in the plants themselves," NASA reported. "Fallen, decomposing leaves and organic matter provide the majority of nutrients, which are rapidly absorbed by plants and trees after entering the soil. But some nutrients, including phosphorus, are washed away by rainfall into streams and rivers, draining from the Amazon basin like a slowly leaking bathtub."

By calculating the volume of the dust clouds and then analyzing the contents of the dust that drifted across the Atlantic, the researchers were then able to estimate how much phosphorus Africa naturally sends the Amazon annually: about twenty-two thousand tons. And the annual tonnage of phosphorus lost from the Amazon due to erosion and flooding? It is, according to the NASA study, about the same amount it receives from Africa.

The study only looked at seven years of dust transport, and the load that moved each year varied, but the phosphoric link between desert and jungle—Africa and South America—plainly left an impression on the scientists. "This is a small world," concluded the study's lead author, Hongbin Yu of the University of Maryland. "And we're all connected together."

In the last two hundred years, humans cracked the circle of life held together by phosphorus and replaced it with a line running straight from mines to farms to waters that are, as a consequence, increasingly fouled by toxic algae.

But there are steps we can take to pull some of that nuisance phosphorus back into the agricultural cycle, which could check burgeoning algae outbreaks *and* extend the life-expectancy of the Earth's phosphorus reserves.

Consider the vast amount of phosphorus lost during its mining, refining, and transport—up to 50 percent. Then consider the phosphorus lost due to erosion before a crop can take it up, let alone the phosphorus squandered in the form of discarded food.

"We waste about 80 percent of the phosphate we use specifically for food production," says Dana Cordell, an Australian food sustainability expert who has done path-breaking research on the future of phosphorus supplies. "It's getting lost at all stages, from mine to field to when we produce and consume food."

And even a significant amount of the phosphorus that *does* take a trip into a corn stalk, and then into a cow and then into our meat and dairy products, is still ultimately bound for rivers, lakes, and oceans, in the form of manure seeping off farm fields, or in our own phosphorus-rich sewage.

These are fixable problems, as is the damage being done by the federal government's ethanol mandate, whose years (at least as it is currently configured) are likely numbered, if for no other reason than the surging electric vehicle sales. But we shouldn't wait for the market to fix the mess made by a product whose most potent renewable feature, it turns out, is the boost it gets every four years from presidential aspirants campaigning in Iowa.

The biofuel lobby has proven so powerful that it even compromised one of the most influential conservationists on the planet—Al Gore, once a full-throated booster for federal ethanol subsidies. "One of the reasons I made that mistake," confessed the author of *An Inconvenient Truth*, "is that I . . . had a certain fondness for the farmers in the state of Iowa because I was about to run for president."

An obvious remedy is to push Iowa back on the presidential campaign calendar—a move the Democratic Party began mulling for other reasons in 2022.

And while the picture that emerges from the chronic polluting, wanton wastefulness, and ill-conceived policy might be distressing, we have seen it before.

IT'S BEEN A half century since pioneering ecologist David Schindler grabbed his 35 mm camera, climbed into a bubble-canopied helicopter, and thundered off into the Canadian wilderness to capture the iconic image of the remote lake that he had intentionally overdosed with phosphorus fertilizer.

The photo of the lake's ghastly green surface served as Schindler's closing argument that phosphorus-packed detergents were behind the algae outbreaks ravaging Lake Erie and freshwaters across the continent during the middle of the twentieth century. Schindler could not have made his case more convincingly, and the public's verdict was swift—fouled waters were too high a price to pay for whiter and brighter shirts and sheets. Phosphorus use in laundry detergents was eliminated or greatly scaled back in the 1970s and '80s, and in a little more than a decade, the nation's algae problem was significantly reduced.

While a more troubling phosphorus picture is emerging today, it is not a hopeless one.

Not long before Schindler died in 2021 at the age of eighty, he warned me that the new, toxic wave of algae outbreaks on waters across the continent—and the globe—aren't going to be as simple to fix this time because the problem has grown in scope and become more complex.

There were, Schindler explained, only a handful of detergent makers in the 1970s that needed to change the way they did business. Today the phosphorus problem in the United States alone is driven by the roughly two million farms operating on about 40 percent of the landmass of the lower forty-eight states. Compounding the menace of the ongoing agricultural pollution, Schindler explained, is all the "legacy" phosphorus tied up in farm soil due to decades of over-fertilizing during an era when farmers were often told by agricultural experts that more-is-better when it came to spreading fertilizer on their fields. Those saturated soils will be leaching that excess phosphorus for years to come. But Schindler insisted this doesn't mean we shouldn't begin to let science again guide us in fixing what he still viewed as a solvable problem.

"What we better be prepared to do is make pretty draconian restrictions on phosphorus applications and on land uses that promote runoff, but then also be very patient," Schindler told me. "This isn't going to happen in a few years. It's that patience factor that always stymies things, because for some reason people can spend fifty years screwing up a lake but expect to be able to fix it in a couple of years. It just isn't the way it happens."

James Elser is a University of Montana ecologist and the director of Arizona State University's Sustainable Phosphorus Alliance. His group works with researchers and the various industries whose futures are tied to phosphorus—fertilizer manufacturers, crop growers, dairy owners, food producers, wastewater treatment operators—to figure out how to re-engineer a more sustainable phosphorus system. Elser, who has coauthored his own excellent book on the subject, predicts the need to change our wasteful, water-polluting ways is on the verge of becoming acute.

"By 2050, nine, ten billion people will need to be fed and they are becoming more affluent as well, which is great, but that also means more meat production, and that puts more pressures on the phosphorus system," he says. "We have to do that at the same time that we don't forget people need to drink water as well. . . . We have to make these two things happen at the same time. That's a challenge."

Elser agreed with Schindler that the original phosphorus mess created—and largely resolved—in the twentieth century pales in comparison to today's challenges. "You've dispersed this phosphorus across the landscape. Now it's in runoff. It's in groundwater and it's blowing off the land in dust," Elser says. "It's everywhere and it's hard to solve that problem, and it's not an activity that you can just stop because you still have to grow food, obviously. So it's a much more difficult problem, but we're on it now!"

One encouraging example: About a dozen states had already moved to ban phosphorus lawn fertilizer before lawn-care giant Scotts Miracle-Gro Company voluntarily pulled it from most of its yard fertilizers over a decade ago. It was a small step because over-fertilized lawns were considered a minor piece of the problem on the global scale (though maybe not on *your* particular lake). But the move was significant in that it signaled a growing public—and fertilizer industry—awareness of the phosphorus-water pollution nexus.

It is an awareness that needs to keep growing, and that will only happen when people wake to the notion that a phosphorus atom isn't nature-designed to be used once and then flushed away.

Take water. All the $H_2O$ we have on Earth now is all we will ever have. Water molecules might get polluted with contaminants for a time, or they might get locked up in glaciers for eons, or whole regions may suffer decades of drought, but Earth's overall water balance never fluctuates. So we're never going to run out of water. That doesn't mean we don't have to worry about supply, both undersupply due to pollution, drought, and water diversion projects or oversupply because of climate change. For example, if all the glaciers were to melt over the course of just a few decades, ocean levels could rise some 230 feet, drowning virtually every coastal city (and many others) on the globe.

The phosphorus cycle functions similarly—the phosphorus atoms the Earth has now are essentially all it will ever have. For billions of years, they leached into the living world as their host rocks eroded, like drops of water off a melting glacier. Now that we have figured out how to turn those drips into a gusher by mining the sedimentary rocks created by dead sea life raining down upon the ocean floor, we are flooding the world with it—in some cases to disastrous effect. Like water, we can't live without phosphorus but, like water, too much of it causes its own set of drastic problems.

"We've taken millions and millions of years of phosphorus that had accumulated in these (sedimentary rock) deposits and released it into the world in the last fifty years . . . and the impacts of that are not over," says Elser. "Phosphate is a biological accelerant, that's what I call it. It's like spraying gasoline on a forest fire. It just makes life go crazy."

To slow the torrent we've unleashed will require a change in our relationship with the element beyond becoming more efficient in how we mine, process, and apply chemical fertilizer.

It means shedding our ingrained notion that human and animal excreta is waste, because it is anything but that.

NORTH KOREAN SOLDIER Oh Chong Song, son of an army general, was drinking and driving on a gray November afternoon in 2017 when he blew past a military checkpoint just north of the DMZ. Once he grasped the gravity of his transgression, the twenty-four-year-old hit the accelerator and raced for the nearby South Korean border, where he dumped his jeep and made a dash for the 38th Parallel as fellow North Korean soldiers opened fire. They riddled Song with so many bullets that he was left for dead in the middle of no-man's-land but, remarkably, he did not bleed out.

Soldiers from the South eventually dragged Song to safety, and he was helicoptered to Seoul, where life-saving surgery revealed his guts were ravaged by more than gunfire. When surgeons opened him up to stitch his shredded bowels, a cluster of nearly foot-long, flesh-colored parasitic worms burst forth.

US journalists were aghast to learn the suspected source of the worms was the human excrement that famished North Koreans were spreading on farm fields as fertilizer, and much was made of a recent North Korean government directive to increase sewage spreading on crops to boost food production for the twenty-five million citizens of the famine-plagued nation.

A *Newsweek* headline proclaimed: KIM JONG UN MAY HAVE CAUSED PARASITIC WORM EPIDEMIC IN NORTH KOREA BY MAKING FARMERS SPREAD HUMAN FECES ON THEIR CROPS.

Supreme Leader Kim Jong Un's government has no doubt done untold horrible things to its people, but directing farmers to use human waste as crop fertilizer isn't necessarily one of them. It's actually common sense. At least it was before the European sewer revolution in the mid-1800s.

Few people in the developed world today bother to ponder what happens to the contents of their toilet flushes. But human excrement was never far from the minds—or nostrils— of nineteenth-century Londoners who were served by an aged, undersized sewer system not up to the task of carrying the burgeoning city's human waste into the River Thames. This meant that much of London's human waste was removed manually from the basements and backyard cesspits into which it was dumped. A brigade of nocturnal laborers known as "night-soil men" did the work with shovels, buckets, and carts. They toiled at night so everyday Londoners didn't have to share the streets with all the leaky carts dripping excrement.

Some of the collected waste was dumped in waterways, but often it was carted out to the countryside to be composted in a manner that baked out many of the dangerous bugs so the excrement could be used as crop fertilizer, a practice once common throughout Europe.

But as London's population ballooned during the Industrial Revolution so, of course, did the volume of its waste, as well as the distance the night-soil men had to travel to get to farm country. By the middle of the 1800s, London had exploded into a metropolis the size of which the Earth had never seen, and its excreta was piling faster than it could be safely hauled away, leaving the city's 2.5 million residents exposed to scourges like cholera.

Most scientists at the time blamed all the pestilence on fouled air, but London physician John Snow came to believe tainted water was the problem, and when a cholera outbreak hit London in 1854, he famously set out to prove his theory. He started by mapping the homes of victims. Then, through exhaustive tracing, he determined that many of them shared one thing in common—water fetched from the Broad Street pump in London's Soho neighborhood.

It turned out that pump was just feet from a leaking cesspit. And it further turned out that a mother, whose baby had been stricken with cholera, had been washing the child's diapers with water she dumped into the cesspit, and it was dripping into the drinking water well. Snow convinced local officials to disable the pump by removing its handle, and soon after the cholera outbreak, already waning, faded away.

Today the Broad Street pump study is recognized as an unqualified success in the history of public health (though on the evening when I visited the replica pump installed on the site of the original, few of the patrons at the adjacent John Snow pub showed any interest in it—other than as a target at which to fling their cigarette butts). Snow's work might have awakened Londoners to the dangers of contaminated water, but it alone wasn't enough to prompt government leaders to make significant changes in the way the city disposed of its sewage.

Then came the "Great Stink" of 1858. During that year's scorching, dry summer, mounds of feces stewing on the banks of the Thames radiated a pong so foul that the curtains of Parliament's riverside Palace of Westminster had to be soaked in disinfectant to dull the stench, and even then lawmakers still shuffled about Westminster's halls with handkerchiefs pressed to their noses.

The miasma focused politicians' attention as no epidemiological study could. They declared war on London's human waste, and a system of train tunnel-sized sewer mains fed by a web of smaller pipes was built to drain city waste into the Thames, where river currents and ocean tides could pull the filth out to sea.

The improvement in the city's air and drinking water quality was immediate, and it was not long before cities across Europe and North America followed London's lead.

Flushing away the Western world's municipal waste in this fashion reduced the disease and putridity that had plagued nineteenth-century urban life, but it came with a heavy set of costs, some anticipated, some not. Not only were sewer systems expensive to build and operate, but piping all that phosphorus-rich poop into rivers and lakes also triggered noxious algae outbreaks in public waterways across Europe and North America.

More significantly, shunting human waste into waterways in this manner permanently cracked the phosphorus circle in a manner that helped to put the Western world on a path to chemical fertilizer addiction.

One of the first to recognize the downside to London's sewer system was fertilizer pioneer Justus von Liebig, who considered disposing of the waste generated by the world's largest city in this manner a colossal economic and agricultural blunder—another step down a misguided path that would inevitably lead to England's nutritional ruin. No matter how many graveyards, dung heaps, and rock deposits England scavenged around the globe, Liebig argued, some day they would eventually play out. All of them.

"The farmer of the present day believes that the introduction of manures from abroad will have no end," Liebig wrote to the *Times* of London in 1859, just as the city embarked on its sewer-building binge. "It is much simpler, he thinks, to buy guano and bones, than to collect their elements from the sewers of cities, and if a lack of the former should ever arise, it will then be time enough to think of a resort to the latter." It was a strategy Liebig called "dangerous and fatal" because eventually fertilizer-exporting countries would run low on their own supplies and would be forced to cut off exports.

But, Liebig argued, harvesting London's sewage on a grand

scale would be one project that would address two pressing problems.

"I am not ignorant of the difficulties which stand in its way—they are indeed very great; but if the engineers would come to an understanding with the men of science in relation to the two purposes—the removal of the contents of the sewers, and the recovery of their valuable elements for agriculture—I do not doubt that a good result would follow," Liebig wrote, adding that if Londoners could not solve this problem, he doubted any European city could.

Nineteenth-century journalist Henry Mayhew viewed the flushing away of human-generated nutrients in starker economic terms, reporting that by the 1850s, English farmers were spending about two million English pounds per year to import foreign fertilizer. At the same time, the city of London was dumping some forty million tons of fertilizer-rich sewage into the Thames each year. That, he calculated, was the equivalent of throwing away nearly 250,000,000 pounds of bread annually.

"By pouring into the river that which, if spread upon our fields, would enable thousands to live," he wrote, "we convert the elements of life and health into the germs of disease and death."

A similar conservation ethic fermented at the same time in learned circles on the other side of the English Channel. Victor Hugo himself observed in 1862 how absurd it was for Europeans to be scraping petrified animal poop from some of the most farthest-flung spots on the globe at the same time European cities were throwing away their homegrown product.

"All the human and animal manure that the world wastes, if it were put back into the land instead of being thrown into the sea, would suffice to feed the world," Hugo observed in *Les Misérables*. "Those heaps of excrement at boundary-posts, those cartloads of muck jolted through the streets at night, those frightful vats at the municipal dumps, those fetid seepings of subterranean sludge that pavements hide from you—do you know what they are? They are the meadow in flower, green grass . . . thyme and sage, they are game, they are cattle . . . they are fragrant hay, golden wheat, they are the bread on your table; they are warm blood in your veins, they are health, they are joy, they are life."

This might have been a radical notion for mid-nineteenth-century Europeans to digest, but it was—and remains to some degree to this day—a fact of life in Asia.


WESTERN AGRICULTURAL AND health experts who visited Asia's biggest cities in the late 1800s were stunned when they witnessed the upside of the "outdated" practice of using night soil as fertilizer. "While the ultra-civilized Western(er) elaborates destructors for the burning of garbage at a financial loss and turns sewage into the sea, the Chinaman uses both for manure," wrote the British health officer for the city of Shanghai in 1899. "He wastes nothing while the sacred duty of agriculture is uppermost in his mind."

By creating a sophisticated network to spread composted human waste across their croplands, the Chinese proved everything Liebig argued. They did not have to plunder other nation's phosphorus reserves. They did not have to finance sewer systems. And they did not have to worry about those sewers contaminating their water supplies.

Nineteenth-century European-style sewer systems (which dumped waste straight into waterways with no sewage treatment at all) would have, the health official reported, led to "sanitary suicide" in the more densely populated East. "And in reality," he wrote, "recent bacterial work has shown that faecal matter and house refuse are best destroyed by returning them to clean soil, where natural purification takes place."

Major cities in Asia refined their process of recycling human waste with sophisticated, closed-loop waste treatment systems that relied on wheels rather the pipes. Farmers carted their grains, vegetables, and animals into the cities, and used the same carts to haul back to the countryside the waste those cities generated. Then they used that waste to grow more grains, vegetables, and animals that they would then cart back to the cities. And on and on it went. *For thousands of years.*

In 1909, just as rock mining was beginning to ravage tiny Banaba Island, pioneering US soil scientist Franklin King took a nine-month tour of China, Korea, and Japan to learn how those countries continued to prosper agriculturally, despite having far more people—and far less agricultural land per person—than the United States.

Americans, he observed, had run into soil fertility issues in just a matter of decades of farming the newly settled continent, whereas Asian agriculturalists, working their same patches of land for centuries, were able to maintain soils that were up to four times as productive as America's increasingly chemically enhanced croplands.

"When we reflect on the depleted fertility of our own older farm lands, comparatively few of which have seen a century's service, and upon the enormous quantity of mineral fertilizers which are being applied annually to them in order to secure paying yields," King wrote in his seminal work, *Farmers of Forty Centuries; Or, Permanent Agriculture in China, Korea and Japan*, "it becomes evident that the time is here when profound consideration should be given to the practices the Mongolian race has maintained through the centuries."

The idea of making poop disappear down a pipe appealed to Westerners in terms of protecting public health and convenience. But what Europe had come to view as noxious waste had, since antiquity, been recognized in many cities across the East as a precious commodity—and not in an abstract way. King reported that in 1908 the rights to collect night soil for a single year in just one section of Shanghai was literally sold for gold—about $1 million worth in today's dollars.

During the extended tour, King's group visited a rice field irrigated with water drawn from a well by a cow yoked to a merry-go-round-like wheel that powered the well pump. King was initially distressed to see that a boy had been tasked with following that cow around and around to capture its dung with a wooden dipper attached to a six-foot bamboo pole, which he deftly tipped to fill buckets with the fertilizer-rich goo.

"There came a flash of resentment that such a task was set for the lad, for we were only beginning to realize what lengths the practice of economy may go, but there was nothing irksome suggested in the boy's face," King reported. "He performed the duty as a matter of course and as we thought it through there was no reason why it should be otherwise. In fact, the only right course was being taken. Conditions would have been worse if the collection had not been made. It made possible more rice. Character of substantial quality was building in the lad which meant thrift in the growing man and continued life for a nation."

King toured Japanese compost houses to see how the excreta was, over a period of five to seven weeks, naturally converted from bacteria-laden municipal and agricultural refuse into a fertilizer trove. He cited a soil analysis that showed Japanese farmers were, by utilizing their waste stream, annually returning to the land as much phosphorus (and nitrogen and potassium) as they were taking.

"These people are now and probably long have been applying quite as much of these three plant food elements to their fields with each planting as are removed with the crop," King concluded. "Moreover there is nothing in American agricultural practice which indicates we shall not ultimately be compelled to do likewise."

Western-style sewage systems are now, of course, common

across urban Asia, but an ethic of sustaining soil productivity instilled over thousands of years dies hard; a 2014 study showed 85 percent of rural homes in five China provinces continued to use household wastes, including sewage, to fertilize crops.

But knowing what we know today about the dangers of microbe-laced human waste, isn't it perilous to be mixing feces with the food supply? I asked Phillip Barak, a professor of soil science at the University of Wisconsin who has an office in King Hall, named after Franklin King himself. Barak answered my question with one of his own: "When is the last time you had raw vegetables at a Chinese restaurant?" Then he told me a story of his own agricultural tour through China as a graduate student in the mid-1980s. One day-trip ended at a field bursting with radishes grown with fertilizer derived from human waste. The allure of it all was apparently too much for a German professor on Barak's field trip. He surreptitiously plucked a single radish from the field and chomped on it, like an on-the-road business-man sneaking an olive from the salad bar at a Ruby Tuesday.

"All our Chinese hosts made disgusted faces," Barak told me, explaining they were aghast because they knew what went on the field to make that radish grow. Even when properly composted, microbes can persist in human waste and that, Barak explained, is why vegetables used in traditional Chinese cuisine are so commonly served cooked. "That," he told me, "was their sanitation."

Barak was happy to talk about the history of human waste-based fertilizer and his work on a project to recover more of it from the local sewage treatment plant, but he also wasn't about to disparage the farming or chemical fertilizer industries that sustain humanity today.

"Agriculture can only keep up with population growth because of the fertilizer industry," he said. "You could say this has caused a problem, that we shouldn't have 7 billion people on the planet. OK, then who gets to be here, and who doesn't?"

Yet Barak acknowledged there are limits to how far humanity can go on mined phosphorus. "The whole agricultural system is predicated on the idea: 'Use what you need, we'll make more,'" he said. This notion leaves him worrying about what is in store for future generations left with a food system that plainly cannot be sustained over the long term.

Mining industry officials maintain that there are enough reserves to last at least another 350 years while, as we've seen, some phosphorus experts contend that dangerously destabilizing regional phosphorus shortages could come in a matter of decades. But even the rosy 350-year horizon does not buy humanity much time. It is, coincidentally, almost exactly the same amount of time between today and the day Hennig Brandt discovered phosphorus in his Hamburg lab back in 1669.

Whatever the exact number of years we have before phosphorus deposits run problematically low for food production in at least some places on the globe, there is no question that we are blowing through the stuff in a manner that will baffle future generations.

"What will they think of us," Barak asked, "squandering and taking all these resources from our children?"

That question got me wondering what *previous* generations would think about what has become of modern agriculture.

And that got me thinking what the young Chinese laborer whose job was to follow a cow around with a wooden dipper think about a modern American dairy with thousands of cows and pond-sized manure lagoons.

He'd probably see riches.

◊　◊　◊

MODERN SLAUGHTERHOUSES ARE surprising models of thrift undertaken on an industrial scale. Beyond the parts of a slaughtered cow used for meat, much of the rest makes its way to market in other ways. Hides become car seats, wallets, shoes, and sofas. Fat is processed into soap, body creams, lipstick, and toothpaste. Organs are used to make medicines like insulin, steroids, and blood thinners. Gelatin derived from boiling bones ends up in marshmallows.

But the "wastes" produced while the cows are *alive* is a different story. Nutrient-rich manure is ripe for similar exploitation. Yet we continue to deal with animal waste in a most medieval manner; we liquefy it so it can be sprayed in a brown mist on open fields, often whether those lands require the nutrient infusion or not.

"At some point people are going to realize that all this manure isn't waste," the chairman of the International Joint Commission, a binational body that oversees US and Canadian boundary water issues (including Lake Erie), once told me. "It's a resource."

We are now at that point.

With farming increasingly conducted on an industrial scale, with herds that number into thousands of head of cattle, with factory farms now capable of producing as much solid waste as a city, a logical next step is for Congress to revisit the agriculture exemption from the Clean Water Act so large-scale farmers are held accountable for being the industrial polluters that they patently have become.

But economics might drive change faster than federal legislation.

In spring 2022, the *Milwaukee Journal Sentinel* reported that, in the heart of America's Dairyland, nearly a dozen farms that collectively manage twenty-five thousand cows were poised to begin pooling their manure so it could be run through a $60 million "digester" that uses bacteria to convert the carbon in the cow waste into methane. That natural gas will then be fed into a network of interstate pipelines that allow the fuel to go anywhere in the country.

The enterprise is driven by a California program that gives petroleum companies tax credits to subsidize low-carbon fuel sources, and manure-derived methane qualifies—even manure as far away as Wisconsin. The *Journal Sentinel* reported the California law has the potential to create something of a manure boom. A 3,500-cow operation could earn as much as $350,000 annually, more if a dairy farmer invests in his own manure digester.

"At that point," the newspaper quoted one industry consultant, "milk has become the byproduct of manure production."

This isn't as upside down as it sounds. The margin for producing milk is at times so slim—or nonexistent—that in 2020 farmers were literally dumping milk on their fields because the demand had plummeted and, like manure, dairy herds don't stop producing milk just because the market goes soft.

The milk oversupply problem is not new; for decades the federal government purchased excess milk supplies and preserved it as a low-grade cheese that it gave away to the poor. The government itself is no longer stockpiling and doling out free cheese in the manner it used to. But federal subsidies mean milk supply, which has hit record levels in recent years, often outstrips demand—so much so that as recently as 2018 some 1.4 *billion* pounds of excess, government-subsidized cheese was stashed in refrigerated warehouses across the country.

So it's not surprising that even as food prices climbed in recent decades, milk prices often went in the opposite direction. One of the reasons milk remains such a relative bargain today, beyond the boost in production and a steep decline in per capita consumption over the past half century, is that farmers aren't paying the true cost of its production.

The public is—in the form of closed swimming beaches and threatened drinking water supplies.

"We make it cheap, just by dumping all the manure in our rivers and lakes," says Steve Carpenter, winner of the prestigious

Stockholm Water Prize and professor emeritus at the University of Wisconsin-Madison's center for limnology.

The burgeoning manure boom is driven by methane but, like a cow carcass, there is more to be mined from it. The next step is to concentrate its nutrients—particularly phosphorus and nitrogen—and put them in a form that makes transporting and applying them more economical and more effective than today's system of driving around giant trucks sloshing with dung and spraying it on any field whose owner is willing to take it.

The potential benefits to better managing manure are staggering, both in protecting water quality and preserving phosphorus rock reserves for future generations.

"If all manures were recycled and returned for [agriculture] production, I think you could displace half of the mined fertilizer," says Elser, director of the Sustainable Phosphorus Alliance. In other words, if we aggressively refined manure for fertilizer, we might essentially double the life of existing phosphorus reserves, based on today's level of phosphorus use.

There also reasons to question our volume of meat production. Nearly one-third of the pork produced in the United States is exported along with nearly one-fifth of our poultry. Do we really want to poison our waters to provide cheap meat abroad, or to produce more milk than anyone wants to buy, and more cheese than anyone wants to eat?

There are, furthermore, ripe opportunities to restore the phosphorus cycle by reclaiming the nutrients in the human waste stream. Globally there is, very roughly, three million tons of phosphorus flowing out the undersides of humans every year in the form of urine and feces, yet relatively little of it is exploited for the fertilizer trove that it is.

◊ ◊ ◊

UNIVERSITY OF MICHIGAN researchers in 2017 cut a ceremonial yellow ribbon on the women's bathroom on the second floor of the university's department of civil and environmental engineering and opened the door to reveal an odd little contraption—a toilet bowl with two drains.

The back drain in the bowl is designed to capture its user's feces, which flush into a sewer pipe that flows to a local wastewater treatment plant on the banks of the Lake Erie-bound Huron River.

The bowl also has a front drain, down which well-aimed urine is intended to flow straight into a tank in a "urine processing room" in the building's basement, where a refrigeration system freezes off the water in the waste and concentrates the targeted nutrients. The men's bathroom is equipped with its own urine-capturing device—a urinal that sends pee straight down a pipe to a tank in the basement.

The toilets are part of a $3 million urine fertilizer research project funded by the National Science Foundation with the dual goals of developing technology to harvest the phosphorus, nitrogen, and potassium from a toilet flush and turn it into a safe fertilizer, and, perhaps more challenging, to sell the public on the idea.

A month after the experiment began, researchers had a modest gallon of concentrated nutrients to show for their efforts. But the potential benefits of scaling up such a urine recovery process are immense, considering most of the phosphorus we excrete is in urine, and considering how much purified drinking water (up to seven gallons) we use to flush away a single pint of pee.

"Our current agriculture system is not sustainable, and the way we address nutrients in our wastewater can be much more efficient," says Krista Wigginton, associate professor of civil and environmental engineering at the University of Michigan and one of the leaders of the research project.

The key is to marry these enterprises, which is exactly what the Michigan researchers are doing in Brattleboro, Vermont, where they are capturing substantially more urine from willing townsfolk and applying it on test plots of carrots, lettuce and wheat.

One of the focuses of the Vermont fieldwork is to experiment with urine treatment methods—including filtering, heating, composting, and evaporating—to make the liquid waste safe for farmers and their customers. Some form of disinfecting treatment is needed because a jug of urine, though far safer than a sack of feces in terms of pathogens, can still harbor bacteria and viruses. Most of them can be naturally neutralized over time. The pharmaceuticals common in today's urine streams are a bigger issue.

"There is no doubt that urine can be a safe fertilizer for growing any kind of crop," says Abraham Noe-Hays, cofounder of the Rich Earth Institute, the Vermont agriculture research enterprise working with the University of Michigan researchers. "The question we're trying to answer is: Is urine from the whole population, taking a whole array of pharmaceuticals, is that safe for unrestricted use in agriculture?"

An equally daunting question: Will consumers buy a bag of carrots raised on pee?

"I think they got a noble effort going, I think they have a lot of work ahead of them—mostly to get public acceptance," says Bruce Lawrence, chief operator of the Brattleboro water treatment facility. "The normal—'normal,' if you will—public person will say: 'Those carrots were grown with pee for fertilizer? No way.' I think that's what's got to be overcome."

Part of the university's grant was spent developing a public relations campaign to do just that. It features a video starring a drop of urine named Uri Nation who advocates for "pee-cycling."

"You probably think of me as waste," Uri says in a singsong Australian accent, "but you've got me all wrong. I'm liquid gold!" The cartoon drop of pee then echoes the argument Victor Hugo made nearly two centuries ago. "Every day the average adult's urine has sufficient nutrients to fertilize enough wheat to make one loaf of bread," Uri boasts. "It seems a pity to waste it."

The Michigan researchers have modeled the costs of diverting urine from sewage treatment plants in terms of energy consumption, freshwater use, greenhouse gas emissions, and algae blooms, and they basically confirmed what Asian farmers have known for centuries, that returning human waste—treated human waste, in this case—to the landscape makes more environmental sense than putting it into pipes and jeopardizing drinking water supplies, swimming beaches, and fishing grounds.

Recycling urine in this manner could be particularly useful in underdeveloped regions across the globe that have yet to invest in extensive wastewater treatment systems. But it gets more complicated for sprawling modern cities with millions of traditional toilets and sewer networks that would have to be radically replumbed to send feces in one direction and urine in another. The researchers acknowledge these challenges but say the opportunities for separating waste streams will come as cities rebuild sewer infrastructure that is, in many places, already well beyond its design life.

TAPPING THE HUMAN *solid* waste stream—feces—that contains far more pathogens than urine is a more daunting challenge. Consider the worms that burst from the North Korean soldier who was reported to have been infected by eating food fertilized with human sewage, or the disgusted look on the Chinese tour guides' faces when the visiting German soil scientist gobbled a raw radish on a field fertilized with feces.

But technologies already exist to refine the nutrients in human waste all the way down to their elemental form.

A wastewater treatment plant in Chicago, for example,

installed a nutrient recovery system several years back that was expected to cut the phosphorus load in its water discharges by about 30 percent. It turns that captured phosphorus into commercial-grade fertilizer pellets—a modest but valuable cache of crop nutrients that would otherwise be flowing toward—and feeding—the dead zone that plagues the Gulf of Mexico.

Some conservationists say Chicago's new system and others like it are a great step, but a true revolution is needed to capture virtually *all* the phosphorus flowing into wastewater treatment plants around the world and convert it back into plant food as safe and contaminant-free as anything produced by a modern fertilizer factory.

That revolution is already happening—in phosphorus's hometown, Hamburg.

Barely two miles from the neighborhood where alchemist Hennig Brandt conjured the first nuggets of elemental phosphorus from a vat of human urine in 1669, a modern-day wizard is again attempting to sift riches from the human waste stream.

Brandt's forays into the mysteries of the natural world were fueled by a lust for gold and navigated by superstition. Martin Lebek works the same waters with a rational mind honed by years of technical studies at Germany's University of Hanover, where he received his doctorate in civil engineering—with a focus on biological wastewater treatment.

I met Lebek in late 2019 at the Hamburg treatment plant that serves more than two million toilet flushers in northern Germany. It is a model of industrial elegance that adds two windmills towering some six hundred feet above the plant's wastewater pools to the Hamburg skyline. Those spinning blades, along with ten onion-shaped, one-hundred-foot-tall waste digester tanks that help convert the methane released from the plant's sewage sludge into energy, together produce enough electricity to power the treatment plant.

Lebek has richer ambitions for Hamburg's sewage. Once that

sewage sludge has been stripped of its methane, it previously met one of two fates. Some was burned and trucked to landfills, and some was spread on agriculture lands to capitalize on its residual phosphorus and other nutrients. The sludge is not technically human waste but the substance produced after carefully cultured bacteria devour the pathogen-rich wastes flowing into the plant.

The treated water that is discharged in the pipes flowing out of treatment plants contain some phosphorus but, by far, most of the phosphorus flowing into a treatment plant ends up in that sludge, also called biosolids.

Enhancing croplands with biosolids is a common practice in Europe as well as the United States. In my hometown of Milwaukee, for example, biosolids are heat-dried into pellets and bagged as a lawn and garden product called Milorganite.

Yet even human waste turned into largely lifeless sludge can still be contaminated with pathogens and other bad stuff—pesticides, pharmaceuticals, heavy metals, and industrial concoctions like those increasingly worrisome "forever chemicals" used in products like nonstick cookware, also known as PFAS. These pollutants may work their way into crops fertilized with biosolids, and thus onto dinner plates and into our bloodstreams. This is why less and less farmland in Europe is fertilized by biosolids. Switzerland has altogether banned the practice, and in Germany today only about a quarter of the biosolids produced by treatment plants make their way back to farm fields. Bigger changes are coming.

Germany will require its largest wastewater treatment plants to strip out basically all the phosphorus contained in its sludge beginning in 2029. The measure passed even while there were questions about whether the technology to accomplish this in a cost-effective way and on an industrial scale could be developed. The private company Lebek works for, Remondis, a family-owned, recycling-focused firm with more than thirty thousand employees, is one among many now in the race to develop such technology.

Remondis started in 2014 with an experimental, small-scale system at the Hamburg wastewater plant that extracted phosphorus atoms from the biosolids after they had been turned into ash. Lebek declined to disclose precisely how his process works because there are a host of other companies competing in the phosphorus recycling rush triggered by the German sewage sludge law. But the essence of the technology is to treat the ash with a precisely measured dose of phosphoric acid in a fashion that releases from that ash more phosphoric acid.

Unlike the super-potent sulfuric acid used to dissolve phosphorus in sedimentary rocks at modern fertilizer factories, Lebek explained phosphoric acid is too weak to release the heavy metals and other contaminants in sewage sludge ash. But it *is* strong enough to let loose the ashes' own phosphoric acid, which is a raw material that can be used to make chemical fertilizer, as well as nutritional supplements in animal feed. Phosphoric acid is also used in human foods, though Lebek said his company has no plans to put its sewage-derived product, pure as it is, into products directly consumed by humans.

The pilot plant process worked so well that by 2019 Hitachi earth movers were roaring in a corner of the Hamburg sewage treatment plant and had already dug the foundation for a full-scale phosphorus recovery facility.

In early 2022, Remondis and its partner in the project, the publicly owned Hamburg water utility, opened the plant and began producing factory-grade fertilizer from the sludge. Lebek anticipated that the plant would be fully operational by the end of the year. He is confident that this recycling technology, applied nationally, could dramatically reduce Germany's reliance on phosphorus imports. This is critical because Europe has no significant phosphorus rock reserves of its own and is therefore just as dependent on foreign fertilizer as the bone- and bird poop–crazy British were in the 1800s. "We're not only recycling phosphorus here to recover a resource," Lebek told me, "but to get independent, mainly from the importing of phosphorus."

And if Remondis or any of its rivals' phosphorus recovery technologies are successfully deployed across Europe, Lebek said it will do more than reduce the continent's dependence on other nations for its food supply. It will also lead to improved water quality, and it should also make some people wealthy along the way.

"We are not crazy to believe that we can make billions of [dollars] here, but this is the first," Lebek said. "And this will be the start of a hopefully longer story."

Lebek knows the story started when phosphorus' elemental powers were unleashed just across the Elbe River more than three centuries ago. Now, less than a century after the city was burned to the ground by Allied bombers dropping phosphorus from the heavens, Hamburg is coaxing from its own ashes a more sustainable food system, and future.

"It is," Lebek said of the phosphorus recovery plant rising along the west bank of the Elbe River, "phosphorus coming home."

# ACKNOWLEDGMENTS

IWROTE MUCH of this book in a Honda minivan in Milwaukee's Lake Park on the western shore of Lake Michigan, and I thank Covid for that.

But before I learned how to work in a van, the pandemic had me leaning harder than usual on my wife Alice and our children: Sarah, Molly, John, and Kate. The 2020 lockdown, in an instant, turned our cramped brick home into a schoolhouse for the kids, an office for Alice, a doghouse for Ernie, the Covid puppy, and a writing "retreat" for me. This left Alice and the kids with little refuge—from each other and from the incessantly yelping dog. I only added to the bedlam, roaming the home in search of a productive place to clack on my keyboard and conduct interviews on the phone, all the while consuming more than my share of the family's bandwidth, Internet and otherwise. So a profound thanks for the support, encouragement and daily accommodating during those trying times goes out to all of them, except the dog.

A big thank-you also goes to the University of Wisconsin-

Milwaukee's School of Freshwater Sciences for providing me financial support (and library privileges) as the journalist in residence at the school's Center for Water Policy.

Literary agent Barney Karpfinger helped shape the outline and focus of this book—not an easy thing to do when you're writing about an element essential to the existence of *every* living cell on the planet. Barney was a tremendous source of support during some exceedingly trying times.

W. W. Norton editor Matt Weiland was a tireless architect (and booster) over the past three years and was more than understanding when family issues required me to put work on hold for weeks on end, sometimes longer. A thank-you as well to Huneeya Siddiqui, Erin Sinesky Lovett, and Steve Colca at Norton for their help in getting the book—and word of it—out.

George Stanley and Marty Kaiser (editor and former editor of the *Milwaukee Journal Sentinel*, respectively) fostered a newsroom culture that allowed enterprising reporters time and space to develop niche beats and pursue complicated stories; without the deep-dive journalism skills I acquired during my two decades at the *Journal Sentinel* this book would not have been possible.

And I never would have considered writing a book about phosphorus had I not encountered an earlier work on the subject by British chemist and author John Emsley, *The Shocking History of Phosphorus: A Biography of the Devil's Element*, which was published in 2000. I came across it while researching phosphorus-driven algae outbreaks on Lake Erie for a 2014 newspaper series. Emsley's book is a great historical overview, packed with riveting accounts of various, often diabolical, applications for phosphorus. But the goal in writing my own book about the Devil's Element is to focus on the paradoxical role that phosphorus plays in our lives today, as an essential crop nutrient and also as the catalyst for all the toxic algae outbreaks raging around the globe. Emsley's book helped set me on that path.

*Acknowledgments*

The work of UK author and fertilizer historian Bernard O'Connor helped me understand the sometimes fuzzy origins of how our modern agricultural system became addicted to chemical fertilizer. Agricultural phosphorus researchers Paul Poulton and Johnny Johnston at the Rothamsted Research center north of London were most helpful during my visit in fall 2019. Michael Paterson, Scott Higgins, and the crew at Canada's Experimental Lakes Area in western Ontario were similarly generous during my visit in spring 2018.

Beyond the expansive cast of characters quoted in this book who shared with me their experiences, expertise and insights, I received invaluable assistance or support from Harvey Bootsma; Val Klump; Joseph Aldstadt; John Janssen; Steve Carpenter; Melissa Scanlan; Jake Vander Zanden; Peter Annin; Boyce Upholt; Cynthia Barnett; Todd Miller; Graceanne Kay Tarsa; Owen Stefaniak; Anna Mayumi Kerber; Larry Boynton; Meg Kissinger; Matthew Mente; Crocker Stephenson; Nancy Quinn, and my eagle-eyed, typo-targeting, octogenarian parents, Dick and Anne Egan.

*AUGUST 10, 2022, LAKE PARK, MILWAUKEE*
*(in my Honda minivan—Covid habits die hard)*

# NOTES

## Introduction

xiv  **"Help, I need help! I'm going to die!!":** *Suspect Nearly Drowns Escaping from Cops* (*The Sun* [UK], September 6, 2018), video, 8:48, accessed April 15, 2022, https://www.youtube.com/watch?v=aJZ-xxRLjdg.

xiv  **smelled like "human feces":** *Washington Post*, September 5, 2018.

xv  **"I need help," said Will Embrey:** Notes from attending meeting in Stuart, Florida, July 26, 2018.

xvi  **declared a public health "crisis":** *Treasure Coast Newspapers*, July 27, 2018.

xvi  **"This is happening!":** Notes from attending meeting in Stuart, Florida, July 26, 2018.

xvi  **"comparable to . . . the great American Dust Bowl":** John R. Vallentyne, *The Algal Bowl: Lakes and Man* (Ottawa: Department of the Environment, Fisheries and Marine Service, 1974), 9.

xvii  **"Be kind to the Earth":** John R. Vallentyne, " 'Johnny Biosphere,' " *Environmental Conservation* 11, no. 4 (1984): 363–64, accessed April 15, 2022, https://www.cambridge.org/core/journals/environmental-conservation/article/johnny-biosphere/DCD355DAB68FF44063A0B91EAD3713B1.

xviii **coroner pointed to a cyanotoxin:** Ian Stewart, Penelope M. Webb, Philip J. Schluter, and Glen R. Shaw, "Recreational and Occupational Field Exposure to Freshwater Cyanobacteria—a Review of Anecdotal and Case Reports, Epidemiological Studies and the Challenges for Epidemiologic Assessment," *Environmental Health: A Global Access Science Source* 5, no. 6 (2006), doi: 10.1186/1476-069X-5-6. There remains some debate about the actual cause of the boy's death.

xix **killing 356 African elephants:** *New York Times*, March 25, 2021.

xix **mollusks will eat almost anything . . . *except* blue-green algae:** Michigan State University, "Are Zebra Mussels Eating or Helping Toxic Algae?," *ScienceDaily* (June 24, 2021), accessed April 15, 2022, https://www.sciencedaily.com/releases/2021/06/210624135534 .htm.

xxi **"accumulations of [fossils] are Fortune's offerings":** *Buffalo Weekly Express*, August 27, 1891.

xxi **fertilizer mines sprawl across . . . central Florida:** "Phosphate," Florida Department of Environmental Protection, Mining and Mitigation Program, accessed April 26, 2022, https://floridadep .gov/water/mining-mitigation/content/phosphate.

xxi **five tons of a mildly radioactive waste material is produced:** "TENORM: Fertilizer and Fertilizer Production Wastes," US Environmental Protection Agency, accessed April 26, 2022, https://www.epa.gov/radiation/tenorm-fertilizer-and-fertilizer -production-wastes.

xxii **men willing to shoot each other over roadbed pebbles:** Arch Fredric Blakey, *The Florida Phosphate Industry: A History of the Development and Use of a Vital Mineral* (Cambridge, MA: Harvard University Press, 1973), 32. (The account may be apocryphal, but as Blakey noted, similar tales were told across central Florida at the time.)

xxiv **"gravest natural resource shortage you've never heard of":** *Foreign Policy*, April 20, 2010.

xxv **Call it the phosphorus paradox:** Tim Lougheed, "Phosphorus Paradox: Scarcity and Overabundance of a Key Nutrient," *Environmental Health Perspectives* 119, no. 5 (2011): A208–13, accessed April 15, 2022, https://doi.org/10.1289/ehp.119-a208.

xxvii **"The environment . . . is all Florida has to offer":** *Naples Daily News*, July 14, 2018.

## Chapter 1: The Devil's Element

4 **"It was like lightning coming out of my jeans":** Interview with author, November 10, 2019.

6 **Nossack recalled just weeks later:** Hans Nossack, *The End* (University of Chicago Press, 2006), 7–8.

8 **develop a more devastating style of urban bombing:** Jörg Friedrich, *The Fire: The Bombing of Germany, 1940–45* (New York: Columbia University Press, 2008), 9.

9 **"marked effect on morale of the enemy":** Arthur Travers Harris, *Bomber Offensive* (London: Collins, 1947), 162.

9 **more than one thousand fortified bunkers:** *New York Times*, October 21, 2019.

9 **the bombs stopped dropping fifty minutes later:** "Royal Air Force Bomber Command 60th Anniversary: Campaign Diary, July 1943."

9 **two-mile-wide whirlwind firestorm:** Jason Forthofer, Kyle Shannon, and Bret Butler, *Investigating Causes of Large Scale Fire Whirls Using Numerical Simulation* (Missoula, MT: USDA Forest Service, Rocky Mountain Research Station, 2009).

9 **winds . . . powerful enough to topple trees three feet in diameter:** John Grehan and Martin Mace, *Bomber Harris: Sir Arthur Harris' Despatch on War Operations, 1942–1945* (Pen & Sword Aviation, 2014), 45.

10 **"an old organ . . . playing all the notes at once":** Igor Primoratz, ed., *Terror from the Sky: The Bombing of German Cities in World War II* (New York: Berghahn, 2010), 98.

10 **Some jumped into canals to snuff the chemical fires:** Eyewitness account from film author viewed in visitor's center at St. Nicholai Cathedral in Hamburg.

10 **weighing piles of ash and estimating:** R. J. Overy, *The Bombers and the Bombed: Allied War over Europe, 1940–1945* (New York: Viking, 2014), 260.

12 **where it likely happened—in Brandt's home laboratory:** Mary Alvira Weeks, *Discovery of the Elements* (Easton, PA: Journal of Chemical Education, 1956), 22.

13 **described . . . as "a man little known, of low birth":** Eduard Farber, *History of Phosphorus* (Washington, DC: Smithsonian Institution Press, 1966), quoting Wilhelm Homberg.

13     **"Hennig Brandt, *Doctor of Medicinea and Philosophiae*":** "Kunckel and the Early History of Phosphorus," *Journal of Chemical Education* (September 1927), 1109.

14     **Turning lead into gold . . . sounds ridiculous today:** This is actually possible to do, though it is fantastically expensive. From Joe Aldstadt, chair of University of Wisconsin-Milwaukee's Department of Chemistry: "In 1980, Glenn Seaborg ripped protons off of Bismuth and made Gold—albeit a few thousand atoms—it cost ~$10,000 to make a billionth of a cent of Gold! So Aristotle was actually (in theory) correct—there is a "prima materia" in that elements are transmutable. I would nominate protons as the prima materia, but a particle physicist would probably differ . . ."

14     **the philosopher's stone . . . could convert . . . lead into pure gold:** Lawrence Principe, *The Secrets of Alchemy* (Chicago: University of Chicago Press, 2013), 125.

15     **"A lot of mischief can come from it":** "Kunckel and the Early History of Phosphorus," *Journal of Chemical Education* (September 1927), 1110.

16     **an eighteenth-century recipe that provides step-by-step instructions:** *Elements of the Origin and Practice of Chymistry*, 5th edition (Edinburgh, 1777), 197–204.

## Chapter 2: The Circle of Life, Broken

19     **"wood, bark and roots had arisen from water alone":** Dmitry Shevela, Lars Olof Björn, and Govindjee, *Photosynthesis: Solar Energy for Life* (Singapore: World Scientific Publishing Company, 2018), 2.

21     **"only a box worth of bones":** Gareth Glover, in discussion with the author, March 18, 2019.

22     **to feed London's hungry denture market:** Bransby Blake Cooper, *The Life of Sir Astley Cooper, Bart., Interspersed with Sketches from His Notebooks of Distinguished Contemporary Characters*, vol. 1 (London: John W. Parker, 1843).

22     **"many of the bones . . . belonged to human beings":** *Morning Post* (London), May 15, 1819.

23    **"a dead soldier is a most valuable article of commerce":** *Morning Post* (London), October 19, 1822.

23    **led to miraculous to crop yields:** *New England Farmer,* February 2, 1827.

23    **"but for the aid of bone manure":** *Chronicle* (Leicester), June 22, 1839.

24    **"Do you wish to be a conqueror?":** Victor Wolfgang Von Hagen, *South America Called Them: Explorations of the Great Naturalists* (New York: Knopf, 1945), 88.

24    **"net-like intricate fabric":** Andrea Wulf, *The Invention of Nature* (New York: Vintage, 2015), 290.

24    **naturalist proclaimed Humboldt "the greatest scientific traveler who ever lived":** Wulf, *Invention of Nature,* 333.

25    **an estimated five million nesting seabirds:** Von Hagen, *South America Called Them,* 154–55.

25    **Andes Mountains suck away moisture:** David Hollet, *More Precious than Gold: The Story of the Peruvian Guano Trade* (Madison, NJ: Fairleigh Dickinson University Press, 2008), 9.

26    **Europe's . . . did not appear to be impressed:** Helmut De Terra, *Humboldt: The Life and Times of Alexander Von Humboldt* (New York: Knopf, 1955), 196.

26    **soon the island had twice as much land in agricultural production:** Gregory T. Cushman, *Guano and the Opening of the Pacific World: A Global Ecological History* (Cambridge University Press, 2013), 30–32.

27    **the rest of his days—all 2,027 of them:** Erica Munkwitz and James L. Swanson, "A Journey to St. Helena, Home of Napoleon's Last Days," *Smithsonian Magazine* (April 2019).

27    **Wellington wrote in 1816 to a friend on St. Helena:** Munkwitz and Swanson, "A Journey to St. Helena."

27    **hundreds of vessels shuttling . . . from the west coast of South America:** R. S. F., "Statistics of Guano," *Journal of the American Geographical and Statistical Society* 1, no. 6 (June 1859): 181–89, https://doi.org/10.2307/196154.

28    **impact . . . was nothing short of "magical":** *Liverpool Mercury,* February 3, 1843.

28    **"The bird is a beautifully arranged chemical laboratory":** Freeman Hunt, "Brief History of Guano," *The Merchants' Magazine and*

*Commercial Review, Vol. 34: From January to June, Inclusive, 1856* (F. Hunt, 1856), 118, reprinted by FB&C, 2017, https://www.google.com/books/edition/The_Merchants_Magazine_and_Commercial_Re/OHpuswEACAAJ?hl=en.

28 **"he felt some of the dust of it going down his throat":** Jimmy Skaggs, *The Great Guano Rush: Entrepreneurs and American Overseas Expansion* (New York: St. Martin's Press, 1994), 6. For a full account, see Charles Kidd, MD, *Medical Times* (J. Angerstein Carfrae, 1845).

29 **a fatality rate greater than 30 percent:** Watt Stewart, *Chinese Bondage in Peru* (Durham, NC: Duke University Press, 1951), 62.

29 **number of laborers shipped to Peru . . . are as high as one hundred thousand:** Benjamin Narvaez, *Coolies in Cuba and Peru: Race, Labor, and Immigration, 1839–1886* (dissertation, University of Texas-Austin, 2010), 4.

29 **miners joined hands and jumped to their deaths:** *Weekly Standard* (Raleigh, NC), June 2, 1858.

30 **continent's guano supply was essentially "limitless":** W. M. Mathew, *The House of Gibbs and the Peruvian Guano Monopoly* (Royal Historical Society, 1981), 146.

30 **nearly twenty-eight billion pounds . . . of guano between 1840 and 1880:** Gregory T. Cushman, "'The Most Valuable Birds in the World': International Conservation Science and the Revival of Peru's Guano Industry, 1909–1965," *Environmental History* 10, no. 3 (July 2005): 477–509.

30 **"anything in short that reminds one of death and the grave":** Alexander James Duffield, *Peru in the Guano Age: Being a Short Account of a Recent Visit to the Guano Deposits, with Some Reflections on the Money They Have Produced and the Uses to which It Has Been Applied* (United Kingdom: R. Bentley and Son, 1877), 89.

30 **After a similarly unsuccessful stint at Oxford:** E. John Russell, *A History of Agricultural Science in Great Britain, 1620–1954* (London: George Allen & Unwin, 1966), 89.

32 **between 1840 and 1880 the average yield . . . nearly doubled:** Yariv Cohen, Holger Kirchmann, and Patrik Enfält, "Management of Phosphorus Resources—Historical Perspective, Principal Problems and Sustainable Solutions," in Sunil Kumar, ed., *Integrated Waste Management*, vol. 2 (London: IntechOpen, 2011), 250.

32 **elements that can be found in abundance in air and water:** Jacek Antonkiewicz and Jan Łabętowicz, "Chemical Innovation in Plant Nutrition in a Historical Continuum from Ancient Greece and Rome until Modern Times," *Chemistry-Didactics-Ecology-Metrology* 21, no. 1–2 (December 2016): 34.

32 **the law of the "limiting factor" was revolutionary:** Fellow German Carl Sprengel is credited by many with pioneering the concepts of mineral-based plant nutrition and the law of the minimum years before Liebig popularized it.

33 **fertilizer prescriptions could be written for individual fields:** William Brock, *Justus Von Liebig: The Chemical Gatekeeper* (Cambridge University Press, 1997), 145.

34 **"repurchase the thousandth part of those conditions of life so frivolously wasted":** Brock, *Justus Von Liebig*, 178.

### *Chapter 3: Bones to Stones*

36 **one of the German soldiers . . . would later recall:** Benjamin A. Hill Jr., "History of Medical Management of Chemical Casualties," in *Medical Aspects of Chemical Warfare*, Textbooks of Military Medicine, ed. Shirley D. Tuorinsky (Washington, DC: US Government Printing Office, August 2014), 80.

37 **chemical attacks . . . would claim 1.3 million casualties on both sides:** Sarah Everts, "A Brief History of Chemical War," Science History Institute (May 11, 2015), accessed April 27, 2022, https://www.sciencehistory.org/distillations/a-brief-history-of-chemical-war.

38 **"made possible by Haber-Bosch nitrogen":** Jan Willem Erisman, Mark A. Sutton, James Galloway, Zbigniew Klimont, and Wilfried Winiwarter, "How a Century of Ammonia Synthesis Changed the World," *Nature Geoscience* 1 (2008): 636–39.

41 **taking a tumble off a nearby bluff:** Patricia Pierce, *"Jurassic Mary: Mary Anning and the Primeval Monsters"* (Gloucestershire, England: The History Press, 2014), 17.

42 **"she understands more of the science than anyone else":** Hugh Torrens, *The British Journal for the History of Science* 28, no. 3 (September 1995): 257–84.

42 **in some coastal areas of England they could be found . . . lit-**

tering the landscape: Larry E. Davis, "Mary Anning: Princess of Palaeontology and Geological Lioness," *The Compass: Earth Science Journal of Sigma Gamma Epsilon* 84, no. 1 (2012): 78.

42 **Buckland described the earthy clumps:** William Buckland, "On the Discovery of Coprolites, or Fossil Faeces, in the Lias at Lyme Regis, and in Other Formations," *Transactions of the Geological Society of London, second series* 3 (1829): 224–25.

43 **"Coprolites form records of warfare":** Buckland, "On the Discovery of Coprolites," 235.

44 **Playfair recalled some years later:** Royal School of Mines (Great Britain), Museum of Practical Geology and Geological Survey, *Records of the School of Mines and of Science Applied to the Arts* 1, pt. 1; *Inaugural and Introductory Lectures to the Course for the Session, 1851–2* (H. M. Stationery Office, 1852), 40–41. I initially encountered this exchange in Bernard O'Connor, *The Origins, Development and Impact on Britain's 19th Century Fertiliser Industry* (Peterborough, England: Fertiliser Manufacturers Association, 1993).

The discovery of rock-based fertilizer cannot be exclusively attributed to Liebig and Buckland (and Anning); other agriculturalists of the era (Lawes included) were engaged in their own hunt for novel forms of fertilizer that also led them to phosphorus-rich rocks.

45 **phosphorus in all that petrified detritus got concentrated:** Stephen M. Jasinski, "Mineral Resource of the Month: Phosphate Rock," *Earth* (January 28, 2015), accessed April 17, 2022, https://www.earthmagazine.org/article/mineral-resource-month-phosphate-rock/.

45 **mining of it peaked in the 1870s:** Harvest data provided to author by fertilizer historian Bernard O'Connor.

45 **harvests had plummeted by the early 1890s:** Trevor D. Ford and Bernard O'Connor, "A Vanished Industry: Coprolite Mining," *Mercian Geologist* 17 (2009), 93–100. (Text provided by author O'Connor.)

46 **harvesting more than one million tons of phosphorus rock annually:** Marc V. Hurst, *Southeastern Geological Society Field Trip Guidebook No. 67: Central Florida Phosphate District*, 3rd edition (Tallahassee, FL: Southeastern Geological Society, July 30, 2016).

46 **"he . . . meant to have it or the red gore would run deep":** Arch

Fredric Blakey, *The Florida Phosphate Industry: A History of the Development and Use of a Vital Mineral* (Cambridge, MA: Harvard University Press, 1973), 32.

47 **"that piece of rock would catch my eye":** Albert F. Ellis, *Ocean Island and Nauru: Their Story* (Sydney, Australia: Angus and Robertson, 1936), 52–53.

47 **Ellis wrote in his diary:** Charlie Mitchell, "New Zealand Can't Shake Its Dangerous Addiction to West Saharan Phosphate," *Stuff*, September 12, 2018.

48 **colleagues offered some harsh words:** Ellis, *Ocean Island and Nauru*, 55. First encountered in Katerina Martina Teaiwa, *Consuming Ocean Island: Stories of People and Phosphate from Banaba* (Bloomington: Indiana University Press, 2014), 43.

49 **Ocean Island, which had been inhabited for at least two thousand years:** Teaiwa, *Consuming Ocean Island*, 48.

49 **The ship's crew found some of the inhabitants:** H. C. Maude and H. E. Maude, eds., *The Book of Banaba, from the Maude and Grimble Papers* (Suva, Fiji: Institute of Pacific Studies, University of the South Pacific, 1994), 72–80. One of the ship's crew members exchanging gifts happened to be an Ocean Island native who had left the island some years earlier and returned with the Australian crew.

49 **As the drought stretched into its third year:** Maude and Maude, eds., *The Book of Banaba*, 83.

50 **Ellis's ship arrived at Ocean Island:** Gregory T. Cushman, *Guano and the Opening of the Pacific World: A Global Ecological History* (Cambridge: Cambridge University Press, 2013), 118.

50 **By the time he finished his hasty survey that first day:** Ellis, *Ocean Island and Nauru*, 58.

50 **In exchange, the Banabans would, collectively, receive 50 pounds per year:** Raobeia Sigrah and Stacey M. King, *Te Rii ni Banaba* (Suva, Fiji: Institute of Pacific Studies, University of the South Pacific, 2001), 170.

51 **The next year, exports ballooned to 13,350 tons:** Ellis, *Ocean Island and Nauru*, 106.

51 **The Banabans were paid less than 10,000 pounds:** Teaiwa, *Consuming Ocean Island*, 18.

51 **the *Sydney Morning Herald* wrote in 1912:** Teaiwa, *Consuming Ocean Island*, 17.

51 **cruel for a people who had recently suffered so immensely:** Pearl Binder, *Treasure Islands: The Trials of the Ocean Islanders* (United Kingdom, Blond and Briggs, 1977), 54.

52 **the *Victoria Daily Times* reported:** *Victoria Daily Times*, July 3, 1920, 21. Author first encountered this description in Cushman, *Guano and the Opening of the Pacific World.*

52 **lots of use for the island's phosphorus but little use for the islanders:** Sigrah and King, *Te Rii ni Banaba*, 329.

52 **the Allies collected the some seven hundred surviving Banabans:** K. J. Panton, *Historical Dictionary of the British Empire* (Rowman & Littlefield, 2015), 384.

52 **last shipments of phosphorus came from the island golf course:** Teaiwa, *Consuming Ocean Island*, 61.

## Chapter 4: War of the Sands

56 **desperately needed natural resources like . . . phosphorus:** Lino Camprubi, "Resource Geopolitics: Cold War Technologies, Global Fertilizers, and the Fate of Western Sahara," *Technology and Culture* 56, no. 3 (2015): 676–703.

57 **the mine employed some 2,600 workers:** Tony Hodges, *Western Sahara: The Roots of a Desert War* (L. Hill, 1983), 127–30.

58 **the United Nations . . . describes Western Sahara as a "Non Self-Governing Territory":** "Security Council Extends Mandate of United Nations Mission for Referendum in Western Sahara, Unanimously Adopting Resolution 2351 (2017)," United Nations, April 28, 2017, accessed April 18, 2022, https://www.un.org/press/en/2017/sc12807.doc.htm.

58 **King Hassan II of Morocco hurried Spain's . . . by sending 350,000 of his subjects across the border:** *Edmonton Journal*, April 9, 1976.

58 **part of Morocco "since the dawn of time":** *Washington Post*, October 21, 2001.

59 **"controls about 80 percent of the world phosphate trade":** *Calgary Herald*, April 9, 1976.

61 **allowing fertilizer production to increase six-fold:** Dana Cordell and Stuart White, "Peak Phosphorus: Clarifying the Key Issues of

a Vigorous Debate about Long-Term Phosphorus Security," *Sustainability* 3, no. 10 (2011): 2027–49.

62 **Earth must double its crop production capacity again:** Deepak K. Ray, Nathaniel D. Mueller, Paul C. West, and Jonathan A. Foley, "Yield Trends Are Insufficient to Double Global Crop Production by 2050," *PLOS ONE* (June 19, 2013), https://doi.org/10.1371/journal.pone.0066428.

62 **scraping from the Earth some 250 million tons of phosphorus rock annually:** "Phosphate Rock," Mineral Commodity Summaries, US Geological Survey, January 2020, accessed April 18, 2022, https://pubs.usgs.gov/periodicals/mcs2020/mcs2020-phosphate.pdf.

63 **food purchases can eat up three-quarters of a family's income:** *New York Times*, April 10, 2008.

63 **there already "is no margin" in the fight against starvation:** "Zoellick Pushes New Approaches for World Bank in CGD Speech," Center for Global Development, April 7, 2008, accessed April 18, 2022, https://www.cgdev.org/article/zoellick-pushes-new-approaches-world-bank-cgd-speech.

64 **"or we will begin to starve":** Jeremy Grantham, "Be Persuasive. Be Brave. Be Arrested (if Necessary)," *Nature* (November 15, 2012).

65 **"reserves of ore are, at any one time, good only for a few decades of use":** Tim Worstall, "What Jeremy Grantham Gets Horribly, Horribly Wrong about Resource Availability," *Forbes* (November 16, 2012).

66 **"once every decade people say we are going to run out of phosphorus":** Renee Cho, "Phosphorus: Essential to Life—Are We Running Out?," Columbia Climate School, April 1, 2013, accessed April 18, 2022, https://blogs.ei.columbia.edu/2013/04/01/phosphorus-essential-to-life-are-we-running-out/.

67 **first page of one recent annual report for the government-run phosphorus fertilizer company:** "Annual Report 2016," OCP Group. Document in author's possession.

67 **"We simply cannot manage for long . . . without Morocco's reserves":** Jeremy Grantham, "The Race of Our Lives Revisited," GMO white paper, August 2018, accessed April 18, 2022, https://www.gmo.com/globalassets/articles/white-paper/2018/jg_morningstar_race-of-our-lives_8-18.pdf.

69 **Najla wrote in 2018 in an open letter:** Najla Mohamedlamin, *Stuff*, September 21, 2018.

## Chapter 5: Dirty Soap

75 **For every Civil War soldier killed on the battlefield, two more died from disease:** J. S. Sartin, "Infectious Diseases during the Civil War: The Triumph of the 'Third Army,'" *Clinical Infectious Diseases* 16, no. 4 (April 1993): 580–84, accessed April 19, 2022, doi: 10.1093/clind/16.4.580.

77 **Procter warned as his researchers began tinkering with various chemical cleaning formulas:** Davis Dyer, Frederick Dalzell, and Rowena Olegario, *Rising Tide: Lessons from 165 Years of Brand Building at Procter & Gamble* (Boston: Harvard Business School Press, 2004), 70. As cited in "Development of Tide Synthetic Detergent," American Chemical Society, 2006, accessed April 19, 2022, https://www.acs.org/content/acs/en/education/whatischemistry/landmarks/tidedetergent.html#inventing-tide.

77 **a "builder" chemical . . . that could neutralize the hard-water minerals:** *The Development of Tide* (booklet), American Chemical Society, October 25, 2006, accessed April 18, 2022, https://www.acs.org/content/dam/acsorg/education/whatischemistry/landmarks/tidedetergent/development-of-tide-commemorative-booklet.pdf.

78 **a box of it could make "oceans of suds":** Davis Dyer, Frederick Dalzell, and Rowena Olegario, *Rising Tide*, 75–76.

78 **by the early 1950s P&G had become the biggest advertiser in the nation:** "Neil McElroy of Procter and Gamble—*Time* Magazine 1953 Article," Marketing Master Insights (blog), April 7, 2012, accessed April 19, 2022, http://marketingmasterinsights.com/input/tag/neil-mcelroy/.

78 **selling a billion pounds of synthetic detergent annually:** *Appleton Post Crescent*, October 24, 1951.

79 **blobs of bubbles . . . so thick they caused car crashes:** *Chicago Tribune*, January 13, 1963.

79 **One bubble mass . . . crested almost five stories above the riverbank:** *Pittsburgh Press*, September 2, 1964.

79    **drew from the same waters into which the suds were discharged:** UPI via *St. Petersburg Times,* July 29, 1962. From the actual article:

## FREE SOAP COMES OUT OF SPIGOT

> Mrs. Raymond Joyce of 14 Eastwood Drive stacked the breakfast dishes next to her sink one morning recently, turned on the water and waited for the suds to build up. She then did the dishes. It happens every day.

80    **Reuss testified before his congressional colleagues upon his return:** *Bristol (PA) Daily Courier,* February 19, 1963.

80    **spokesman . . . told a group of sanitation engineers at a . . . symposium on the bubble conundrum:** *Minneapolis Star,* December 8, 1962.

82    **newspaper columnists at the time wrote Lake Erie eulogies:** *Oil City Derrick,* March 31, 1966.

82    **Added an editor from Ohio:** *Times Recorder* (Zanesville, Ohio), April 15, 1967.

82    **amount of dissolved phosphorus in Lake Erie had nearly tripled:** Rep. No. 91-1004, *Phosphates in Detergents and the Eutrophication of America's Waters,* 91st Congressional Session (April 14, 1970), 6.

83    **churning out some four billion pounds of detergent annually:** A. H. Phelps Jr., "Air Pollution Aspects of Soap and Detergent Manufacture," *Journal of the Air Pollution Control Association* 17, no. 8 (1967): 505–7, doi: 10.1080/00022470.1967.10469009.

83    **as much as 70 percent of the phosphorus in wastewater could be traced back:** Rep. No. 91-1004, *Phosphates in Detergents and the Eutrophication of America's Waters,* 73.

83    **detergent . . . percent phosphate by weight:** David Zwick, Marcy Benstock, and Ralph Nader, *Water Wasteland: Ralph Nader's Study Group Report on Water Pollution* (New York: Grossman, 1971), 451.

83    **spokesman testified at a 1969 congressional hearing:** Rep. No. 91-1004, *Phosphates in Detergents and the Eutrophication of America's Waters,* 63–64.

84    **"Eutrophied lakes may be a small price to pay":** Rep. No. 91-1004, *Phosphates in Detergents and the Eutrophication of America's Waters,* 29.

85 **"easier to do something about three than about a couple hundred million":** Rep. No. 91-1004, *Phosphates in Detergents and the Eutrophication of America's Waters*, 49.

85 **detergent industry, meanwhile, argued there was "no evidence":** Rep. No. 91-1004, *Phosphates in Detergents and the Eutrophication of America's Waters*, 21.

87 **the scholarship committee started to pepper him with unexpected questions:** *Star Tribune* (Minneapolis, MN), December 24, 1961.

88 **key to understanding the way energy flows through a lake:** Nick Zagorski, "Profile of David W. Schindler," *Proceedings of the National Academy of Sciences* 103, no. 19 (May 9, 2006): 7207–9, accessed April 19, 2022, http://www.pnas.org/content/103/19/7207#ref -3.

93 **"Although the results of the Lake 227 experiment silenced those":** D. W. Schindler, "A Personal History of the Experimental Lakes Project," *Canadian Journal of Fisheries and Aquatic Sciences* 66, no. 11 (October 22, 2009): 1140, https://doi.org/10.1139/F09-134.

95 **"We may live out the rest of our lives in a world of garbage":** *Boston Globe*, July 21, 1970.

95 **the US detergent industry agreed to limit phosphorus content:** David W. Litke, *Review of Phosphorus Control Measures in the United States and Their Effects on Water Quality*, US Geological Survey Water Resources Investigations Report 99-4007 (1999), 5, accessed April 20, 2022, https://pubs.usgs.gov/wri/wri994007/ pdf/wri99-4007.pdf.

95 **ban . . . led to an unsuccessful lawsuit by the detergent industry:** *Nanaimo Daily News*, June 9, 1971.

95 **industry voluntarily pulled phosphorus from household detergents:** Litke, *Review of Phosphorus Control Measures*, 1.

## Chapter 6: Toxic Water

98 **"I needed to be by that lake":** Interview with author, July 9, 2018.

99 **a ballooning number of livestock operations:** "GLRI FA3 Priority Watershed Profile: Maumee Watershed," Great Lakes Commission, accessed April 19, 2022, https://www.glc.org/wp-content/ uploads/Maumee-Watershed-Profile.pdf.

99    **Agriculture is responsible for the rest:** "Lake Erie Phosphorus-Reduction Targets Challenging but Achievable," *Michigan News*, University of Michigan, accessed April 19, 2022, https://news.umich.edu/lake-erie-phosphorus-reduction-targets-challenging-but-achievable/.

102    **farm . . . can qualify to keep its manure disposal operations largely unregulated:** Ohio EPA, "CAFO NPDES Permit–General Overview of Federal Regulations," Ohio Environmental Protection Agency factsheet, accessed April 19, 2022, https://epa.ohio.gov/static/Portals/35/cafo/NPDESPartI.pdf.

102    **standards for how much indoor space is required for each type of farm animal:** "Explosion of Unregulated Factory Farms in Maumee Watershed Fuels Lake Erie's Toxic Blooms," Environmental Working Group, accessed April 19, 2022, https://www.ewg.org/interactive-maps/2019_maumee/.

103    **Myers told me, casting playful conspiratorial glances from side to side:** Interview with author, July 9, 2019.

105    **Australian chemist reported a lake . . . with a "scum like green oil paint":** *Nature* (May 2, 1878): 12.

106    **their blood-purifying organs had turned black as coal:** Ian Stewart, Alan A. Seawright, and Glen R. Shaw, "Cyanobacterial Poisoning in Livestock, Wild Mammals and Birds—an Overview," in H. Kenneth Hudnell, ed., *Cyanobacterial Harmful Algal Blooms: State of the Science and Research Needs*, Advances in Experimental Medicine and Biology, vol. 619 (New York: Springer, 2008), 615–16.

107    **"our legislature is a wholly-owned subsidiary of the farm bureau":** *Toledo Blade*, May 2, 2018.

109    **"We're not actually really growing food for humans there right now":** Interview with author, July 9, 2019.

111    **a different story about 450 miles to the northwest on Wisconsin's Green Bay:** Some of the material in this chapter section originally appeared in a series the author wrote for the *Milwaukee Journal Sentinel* in summer 2014.

112    **rapidly suburbanizing county in the heart of "America's Dairyland":** *Milwaukee Journal Sentinel*, December 13, 2019.

112    **home to some 125,000 . . . squeezed onto roughly 190,000 agricultural acres:** "2017 Census of Agriculture County Profile: Brown County, Wisconsin," US Department of Agriculture,

accessed April 20, 2022, https://www.nass.usda.gov/Publications/AgCensus/2017/Online_Resources/County_Profiles/Wisconsin/cp55009.pdf.

112 **biologist who investigated one of Green Bay's massive fish die-offs:** *Milwaukee Journal Sentinel,* September 13, 2014, accessed April 20, 2022, https://www.jsonline.com/in-depth/archives/2021/09/02/dead-zones-haunt-green-bay-manure-fuels-algae-blooms/8100840002/.

114 **phosphorus discharged annually into the Fox River:** *Milwaukee Journal Sentinel,* September 13, 2014.

114 **upgrading their treatment system . . . could cost about $100 million:** *Milwaukee Journal Sentinel,* December 13, 2019.

114 **"if we're not wise, we could see no water quality improvement":** *Milwaukee Journal Sentinel,* September 13, 2014.

115 **sewerage district employee told me of the plan to pay farmers not to pollute:** *Milwaukee Journal Sentinel,* September 13, 2014.

115 **"You could probably enlighten me more than I could enlighten you":** *Milwaukee Journal Sentinel,* September 13, 2014.

117 **Now, when July hits, swimming is just over:** Interview with author, August 7, 2019.

117 **algae blooms had worsened since the 1980s on nearly 70 percent of water bodies:** Jeff C. Ho, Anna M. Michalak, and Nima Pahlevan, "Widespread Global Increase in Intense Lake Phytoplankton Blooms since the 1980s," *Nature* 574 (October 2019): 667–68.

117 **annual blooms are already costing the United States more than $4 billion:** Ho, Michalak, and Pahlevan, "Widespread Global Increase in Intense Lake Phytoplankton Blooms," 667–70.

*Chapter 7: Empty Beaches*

121 **It includes an annual load of roughly 1.6 million tons of nitrogen:** *Mississippi River/Gulf of Mexico Watershed Nutrient Task Force 2019-2021 Report to Congress, US Environmental Protection Agency,* U.S. Environmental Protection Agency, 2022.

121 **Iowa . . . has become *the* battleground to reduce the Gulf's dead zone:** *Des Moines Register,* June 22, 2018.

126 **Jubilee . . . documented for more than a century:** "Sporadic

Mass Shoreward Migrations of Demersal Fish and Crustaceans in Mobile Bay, Alabama," *Ecology* 41, no. 2 (April 1960): 292–98.

127 **fish, shrimp, and crabs . . . are perfectly safe to eat:** "Jubilee Occurring in Mississippi Sound; Seafood Safe to Eat, but People Should Use Caution," Mississippi Department of Marine Resources press release, July 27, 2017, https://dmr.ms.gov/jubilee -occurring-in-mississippi-sound-seafood-safe-to-eat-but-people -should-use-caution/.

127 **"It's a toxic algal bloom. Don't eat any of those fish":** Emily Cotton, in discussion with the author, July 24, 2019.

## Chapter 8: A Liquid Heart, Diseased

136 **But you can still read the initials of the people buried beneath:** Author's 2018 visit to cemetery to locate gravestones of 1926 flood victims.

137 **grocery store that collapsed in the swirling current:** *Tampa Tribune*, September 23, 1926, page 4.

137 **Florida spent some $15 million in the 1910s on a network of canals:** *Appleton Post Crescent*, February 9, 1929.

138 **local newspaper editor who argued for construction of a taller and thicker new dike:** *St. Petersburg Times*, September 26, 1926.

138 **"too gruesome for a newspaper story":** *Miami News*, September 23, 1928.

138 **Herbert Hoover . . . promised the survivors federal help was on the way:** Michael Grunwald, *The Swamp* (New York: Simon & Schuster, 2006), 198.

139 **brayed the narrator of the film the Army Corps titled** *Waters of Destiny*: *Waters of Destiny* (US Army Corps of Engineers, ca. 1957), documentary film, 25:50, Florida Memory, State Library and Archives of Florida, https://www.floridamemory.com/items/show/232410.

140 **concentrations of the nutrient in the lake roughly doubled:** Examination of Basin Phosphorus Issues Associated with Lake Okeechobee Watershed Dairies; National Audubon Society. Document in author's possession.

140 **annual load of phosphorus . . . can be as high as 2.3 million pounds:** Joyce Zhang, Zach Welch, and Paul Jones, "Chapter 8B: Lake Okeechobee Watershed Annual Report," in *The South Flor-*

*ida Environment,* 2020 South Florida Environmental Report vol. 1 (West Palm Beach, FL: South Florida Water Management District, 2020), 8B-2, accessed April 21, 2022, https://apps.sfwmd .gov/sfwmd/SFER/2020_sfer_final/v1/chapters/v1_ch8b.pdf.

140 **about ten times what biologists estimate Lake Okeechobee can safely absorb:** "Appendix A: Northern Everglades and Estuaries Protection Program (NEEPP) BMAPs," in *Florida Statewide Annual Report on Total Maximum Daily Loads, Basin Management Action Plans, Minimum Flows or Minimum Water Levels and Recovery or Prevention Strategies* (West Palm Beach, FL: South Florida Water Management District, June 2018), 17, accessed April 21, 2022, https://floridadep.gov/sites/default/files/2_3_2017STAR_ AppendixA_NEEPP.pdf.

142 **the Army Corps acknowledged in one report, "It leaks":** US Army Corps of Engineers, *Lake Okeechobee and the Herbert Hoover Dike: A Summary of the Engineering Evaluation of the Seepage and Stability Problems at the Herbert Hoover Dike* (Jacksonville, FL: US Army Corps of Engineers Jacksonville District, n.d.), accessed May 03, 2022, http://cdnassets.hw.net/15/5a/ f2357d1240f69f864e55df7b18dd/lakeoandhhdike.pdf.

142 **experts for Lloyds of London toured the Hoover Dike after Hurricane Katrina:** Lloyd's Emerging Risks Team, *The Herbert Hoover Dike: A Discussion of the Vulnerability of Lake Okeechobee to Levee Failure; Cause, Effect and the Future* (London: Lloyd's, n.d.), accessed April 21, 2022, https://assets.lloyds.com/media/528d8f9c-c805 -4b60-a592-847b44201bd3/Lake_Okeechobee_Report.pdf.

143 **"the potential for human suffering and loss would be significant":** US Army Corps of Engineers, *Lake Okeechobee and the Herbert Hoover Dike.*

144 **the lake can rise as much as four feet in a month:** Paul Gray, "High Water Levels Threaten the Health of Lake Okeechobee," National Audubon Society, October 24, 2017, accessed April 21, 2022, https://fl.audubon.org/news/high-water-levels-threaten -health-lake-okeechobee.

147 **red tides can start up to forty miles from shore:** National Centers for Coastal Ocean Science, "What Powers Florida Red Tides?," National Oceanic and Atmospheric Administration, November 18, 2014, accessed April 26, 2022, https://coastalscience.noaa .gov/news/powers-florida-red-tides/.

154 **islanders suffered from the disease at a rate up to one hundred times that of what would be expected:** Jonathan Weiner, "The Tangle," *New Yorker*, April 3, 2005.

## Chapter 9: Waste Not

162 **some of the Earth's phosphorus may have been delivered by meteors:** Keith Cooper, "Did Meteorites Bring Life's Phosphorus to Earth?," NASA Astrobiology Program, accessed April 21, 2022, https://astrobiology.nasa.gov/news/did-meteorites-bring-lifes-phosphorus-to-earth/.

162 **"draining from the Amazon basin like a slowly leaking bathtub":** Ellen Gray, "NASA Satellite Reveals How Much Saharan Dust Feeds Amazon's Plants," NASA Earth Science News Team, February 22, 2015, accessed April 21, 2022, https://www.nasa.gov/content/goddard/nasa-satellite-reveals-how-much-saharan-dust-feeds-amazon-s-plants.

163 **vast amount of phosphorus lost during its mining, refining, and transport:** Dana Cordell and Stuart White, "Sustainable Phosphorus Measures: Strategies and Technologies for Achieving Phosphorus Security," *Agronomy* 3 (2013): 86–116.

163 **"We waste about 80 percent of the phosphate":** *Washington Post*, February 16, 2016.

164 **Al Gore, once a full-throated booster for federal ethanol subsidies:** Gerard Wynn, "U.S. Corn Ethanol 'Was Not a Good Policy': Gore," Reuters, November 22, 2010, accessed April 21, 2022, https://www.reuters.com/article/us-ethanol-gore/u-s-corn-ethanol-was-not-a-good-policy-gore-idUSTRE6AL3CN20101122.

165 **the phosphorus problem . . . is driven by the roughly two million farms:** National Agricultural Statistics Service, "2012 Census of Agriculture Highlights: Farms and Farmland," US Department of Agriculture, September 2014, accessed April 22, 2022, https://www.nass.usda.gov/Publications/Highlights/2014/Highlights_Farms_and_Farmland.pdf.

166 **Elser, who has coauthored his own excellent book:** Jim Elser and Phil Haygarth, *Phosphorus: Past and Future* (Oxford University Press, 2020).

166  **"We have to make these two things happen at the same time":** Jim Elser and Sally Rockey, *Phosphorus Forum 2018* (Sustainable Phosphorus Alliance, April 2, 2018), video, 59:11, accessed April 21, 2022, https://www.youtube.com/watch?v=8A9NFkSwji8.

167  **"We've taken millions and millions of years of phosphorus":** James Elser, in discussion with the author, August 3, 2020.

171  **Liebig wrote to the *Times* of London in 1859:** Justus von Liebig, "On English Farming and Sewers," *Monthly Review* 70, no. 3 (July–August 2018), accessed April 2022, https://monthlyreview .org/2018/07/01/on-english-farming-and-sewers/.

172  **"convert the elements of life and health into the germs of disease and death":** Henry Mayhew, *London Labour and the London Poor* (London: Penguin Classics, 2006), 181–82. Passage first encountered in Stephen Johnson, *The Ghost Map* (New York: Riverhead, 2006), 116.

172  **Hugo observed in *Les Miserables*:** Victor Hugo, *Les Miserables*, trans. Christine Donougher (New York: Penguin, 2013), 1126–27.

173  **wrote the British health officer for the city of Shanghai in 1899:** Dr. Arthur Stanley, 1899 annual report, excerpted in F. H. King, *Farmers of Forty Centuries, or Permanent Agriculture in China, Korea and Japan* (Madison, WI: Mrs. F. H. King, 1911), 198–99.

173  **European-style sewer systems . . . would have . . . led to "sanitary suicide":** Stanley, in King, *Farmers of Forty Centuries.*

174  **profound consideration should be given to the practices:** King, *Farmers of Forty Centuries*, 193.

175  **about $1 million worth in today's dollars:** King, *Farmers of Forty Centuries*, 9.

175  **"there was nothing irksome suggested in the boy's face":** King, *Farmers of Forty Centuries*, 201–2.

175  **"nothing . . . indicates we shall not ultimately be compelled to do likewise":** King, *Farmers of Forty Centuries*, 215.

176  **provinces continued to use household wastes . . . to fertilize crops:** Ying Liu, Jikun Huang, and Precious Zikhali, "Use of Human Excreta as Manure in Rural China," *Journal of Integrative Agriculture* 13 (2014): 434–42.

179  **"At that point":** Rick Barrett, *Milwaukee Journal Sentinel*, February 28, 2022.

179  **Steve Carpenter, winner of the prestigious Stockholm Water Prize:** Steve Carpenter, in discussion with the author, August 07, 2019.

180 **Nearly one-third of the pork produced in the United States is exported:** US Meat Export Federation, "U.S. Pork Exports Soared to New Value, Volume Records in 2019," National Hog Farmer, February 06, 2020, accessed April 24, 2022, https://www.nationalhogfarmer .com/marketing/us-pork-exports-soared-new-value -volume-records-2019.

180 **along with nearly one-fifth of our poultry:** Economic Research Service, "Poultry & Eggs," US Department of Agriculture, last updated April 28, 2022, accessed May 24, 2022, https://www.ers .usda.gov/topics/animal-products/poultry-eggs/.

181 **says Krista Wigginton ... one of the leaders of the research project:** University of Michigan news release, January 22, 2020.

182 **whole population, taking a whole array of pharmaceuticals, is that safe:** *Peecycling* (University of Michigan, April 7, 2015), video, 10:08, https://www.youtube.com/watch?v=dCV 3kWhjfI4&t=108s, in Nicole Casal Moore, "A $3M Grant to Turn Urine into Food Crop Fertilizer," University of Michigan news release, September 8, 2016, https://news.umich .edu/a-3m-grant-to-turn-urine-into-food-crop-fertilizer/.

182 **"'Those carrots were grown with pee for fertilizer? No way.'":** *Peecycling* (University of Michigan, April 7, 2015).

182 **a public relations campaign ... for "pee-cycling":** *Uri Nation Introduces Urine Diversion and Urine Derived Fertilizers!* (University of Michigan, September 29, 2018), video, 6:33, accessed April 22, 2022, https://www.youtube.com/watch?v=iX1F4dYLF84&t=4s.

183 **jeopardizing drinking water supplies, swimming beaches, and fishing grounds:** Jim Erickson, "'Peecycling' Payoff: Urine Diversion Shows Multiple Environmental Benefits when Used at City Scale," University of Michigan news release, December 15, 2020, https://news.umich.edu/peecycling-payoff-urine-diversion -shows-multiple-environmental-benefits-when-used-at-city-scale/.

184 **a nutrient recovery system ... expected to cut the phosphorus load in its water discharges:** *Chicago Tribune*, May 15, 2016.

184 **together produce enough electricity to power the treatment plant:** "Energy Transition in the Port: An Economic Success Story," Hamburg Marketing, Germany, 2018, https://marketing .hamburg.de/energy-transition-in-hamburgs-port.html.

# PARTIAL BIBLIOGRAPHY

Ashley, K., D. Cordell, and D. Mavinic. 2011. "A Brief History of Phosphorus: From the Philosopher's Stone to Nutrient Recovery and Reuse." *Chemosphere* 84, no. 6.

Asimov, Isaac. 1974. *Asimov on Chemistry.* Garden City, NY: Doubleday.

Binder, Pearl. 1977. *Treasure Islands: The Trials of the Ocean Islanders.* London: Blond and Briggs.

Blakey, Arch Fredric. 1973. *The Florida Phosphate Industry: A History of the Development and Use of a Vital Mineral.* Cambridge, Mass: Wertheim Committee, Harvard University, distributed by Harvard University Press.

Boerhaave, Herman. 1735. *Elements of Chemistry: Being the Annual Lectures of Hermann Boerhaave. M.D.* Translated from the original Latin by Timothy Dallowe. 2 vols. London: J. and J. Pemberton.

Botting, Douglas. 1973. *Humboldt and the Cosmos.* London: Joseph.

Brock, William H. 1997. *Justus von Liebig: the Chemical Gatekeeper.* Cambridge: Cambridge University Press.

Cordell, Dana, and Stuart White. 2014. "Life's Bottleneck: Sustaining the World's Phosphorus for a Food Secure Future." *Annual Review of Environment and Resources* 39, no. 1 (October): 161–88.

Cushman, Gregory T. 2013. *Guano and the Opening of the Pacific World: A Global Ecological History.* Cambridge: Cambridge University Press.

Dyer, Davis, Frederick Dalzell, and Rowena Olegario. 2004. *Rising Tide: Lessons from 165 Years of Brand Building at Procter & Gamble.* Boston, Mass: Harvard Business School Press.

Dyer, Gwynne. 1985. *War.* First edition. New York: Crown.

Egan, Dan. 2017. *The Death and Life of the Great Lakes.* New York: W. W. Norton.

Ellis, Albert F. 1936. *Ocean Island and Nauru: Their Story.* Sydney, Australia: Angus and Robertson.

Elser, James J., and Philip M. Haygarth. 2021. *Phosphorus: Past and Future.* New York: Oxford University Press.

Emsley, John. 2000. *The Shocking History of Phosphorus: A Biography of the Devil's Element.* London: Macmillan.

*Eutrophication: Causes, Consequences, Correctives; Proceedings of a Symposium.* 1969. Washington, DC: National Academy of Sciences.

Friedrich, Jörg. 2006. *The Fire: The Bombing of Germany, 1940–1945.* Translated by Allison Brown. New York: Columbia University Press.

Grunwald, Michael. 2007. *The Swamp: The Everglades, Florida, and the Politics of Paradise.* First Simon & Schuster paperback edition. New York: Simon & Schuster.

Harris, Arthur Travers. 1947. *Bomber Offensive.* London: Collins.

Henderson-Sellers, Brian, and H. R. Markland. 1987. *Decaying Lakes: The Origin and Control of Cultural Eutrophication.* Chichester, West Sussex: Wiley.

Hodges, Tony. 1983. *Western Sahara: The Roots of a Desert War.* Westport, Conn: L. Hill.

Hollett, D. 2008. *More Precious than Gold: The Story of the Peruvian Guano Trade.* Madison, NJ: Fairleigh Dickinson University Press.

Hugo, Victor, and Christine Donougher. 2015. *Les Misérables.* Translated with notes by Christine Donougher. Introduction by Robert Tombs. New York: Penguin.

Jensen, Erik. 2005. *Western Sahara: Anatomy of a Stalemate.* Boulder, Colo: Lynne Rienner.

Johnson, Steven. 2006. *The Ghost Map: The Story of London's Most Terrifying Epidemic—and How It Changed Science, Cities, and the Modern World.* New York: Riverhead.

Kassinger, Ruth. 2019. *Slime: How Algae Created Us, Plague Us, and Just Might Save Us.* Boston: Houghton Mifflin Harcourt.

Keegan, John. 1976. *The Face of Battle.* London: J. Cape.

King, F. H. 1973. *Farmers of Forty Centuries; or, Permanent Agriculture in China, Korea, and Japan.* Emmaus, Pa: Rodale Press.

Macdonald, Barrie. 1982. *Cinderellas of the Empire: Towards a History of Kiribati and Tuvalu.* Canberra: Australian National University Press.

Macfarlane, Alan. 1997. *The Savage Wars of Peace: England, Japan and the Malthusian Trap.* Oxford: Blackwell.

Mathew, W. M. 1981. *The House of Gibbs and the Peruvian Guano Monopoly.* London: Royal Historical Society.

Maude, H. C., and H. E. Maude. 1994. *The Book of Banaba.* Suva: University of the South Pacific.

Middlebrook, Martin. 1981. *The Battle of Hamburg: Allied Bomber Forces against a German City in 1943.* New York: Scribner's.

Musgrove, Gordon. 1981. *Operation Gomorrah: The Hamburg Firestorm Raids.* London: Jane's.

Nossack, Hans Erich. 2004. *The End: Hamburg 1943.* Chicago: University of Chicago Press.

O'Connor, Bernard, and Leyre Solano. 2014. *The Spanish Phosphateers: The Origins and Development of Spain's Phosphate Industry.* Lulu.com.

O'Connor, Bernard. 1993. *The Origins, Development and Impact of Britain's 19th Century Fertilizer Industry.* Peterborough: Fertilizer Manufacturers Association.

Overy, R. J. 2014. *The Bombers and the Bombed: Allied Air War over Europe 1940–1945.* New York: Viking.

Pierce, Patricia. 2006. *Jurassic Mary: Mary Anning and the Primeval Monsters*. Stroud, Gloucestershire: The History Press.

Principe, Lawrence. 2013. *The Secrets of Alchemy*. Chicago: University of Chicago Press.

Rhodes, Richard. 1986. *The Making of the Atomic Bomb*. New York: Simon & Schuster.

Rosen, Julia. 2021. "Humanity Is Flushing Away One of Life's Essential Elements." *The Atlantic* (February 8).

Russell, Edward J. 1966. *A History of Agricultural Science in Great Britain, 1620–1954*. London: Allen & Unwin.

Sachs, Aaron. 2006. *The Humboldt Current: Nineteenth-Century Exploration and the Roots of American Environmentalism*. New York: Viking.

Salzberg, Hugh W. 1991. *From Caveman to Chemist: Circumstances and Achievements*. Washington, DC: American Chemical Society.

San Martín, Pablo. 2010. *Western Sahara: The Refugee Nation*. First edition. Cardiff: University of Wales Press.

Schindler, David W., and John R. Vallentyne. 2008. *The Algal Bowl: Overfertilization of the World's Freshwaters and Estuaries*. Revised and expanded edition. Edmonton: University of Alberta Press.

Schreiber, Gerhard, Klaus A. Maier, P. S. Falla, and Wilhelm Deist. 1990. *Germany and the Second World War*. Oxford: Clarendon Press.

Shelley, Toby. 2004. *Endgame in the Western Sahara: What Future for Africa's Last Colony?* London: Zed Books.

Shennan, Jennifer, and Makin Corrie Tekenimatang. 2005. *One and a Half Pacific Islands: Stories the Banaban People Tell of Themselves / Teuana Ao Teiterana n Aba n Te Betebeke : I-Banaba Aika a Karakin Oin Rongorongoia*. Wellington, NZ: Victoria University Press.

Sigrah, Raobeia Ken, and Stacey M. King. 2001. *Te Rii ni Banaba*. Suva: University of the South Pacific.

Skaggs, Jimmy M. 1994. *The Great Guano Rush: Entrepreneurs and American Overseas Expansion*. New York: St. Martin's Press.

Stewart, Watt. 1951. *Chinese Bondage in Peru*. Durham, NC: Duke University Press.

Swasy, Alecia. 1993. *Soap Opera: The Inside Story of Procter & Gamble*. First edition. New York: Times Books.

Teaiwa, Katerina Martina. 2014. *Consuming Ocean Island: Stories of People and Phosphate from Banaba*. Bloomington: Indiana University Press.

Threlfall, Richard E. 1951. *The Story of 100 Years of Phosphorus Making, 1851–1951*. Oldbury, England: Albright & Wilson.

Vallentyne, John R. 1974. *The Algal Bowl: Lakes and Man*. Ottawa: Department of the Environment, Fisheries and Marine Service.

Von Hagen, Victor Wolfgang. 1945. *South America Called Them; Explorations of the Great Naturalists: La Condamine, Humboldt, Darwin, Spruce*. New York: Knopf.

Waring, R. H., G. B. Steventon, and Steve Mitchell. 2002. *Molecules of Death*. London: Imperial College Press.

Weeks, Mary Elvira, and Henry Marshall Leicester. 1968. *Discovery of the Elements. Completely Rev. and New Material Added by Henry M. Leicester. Illus. Collected by F. B. Dains*. Seventh edition. Easton, Pa: Journal of Chemical Education.

Williams, Maslyn, and Barrie Macdonald. 1985. *The Phosphateers: A History of the British Phosphate Commissioners and the Christmas Island Phosphate Commission*. Carlton, Vic: Melbourne University Press.

Wulf, Andrea. 2015. *The Invention of Nature: Alexander von Humboldt's New World*. First American edition. New York: Knopf.

Wyant, Karl A., Jessica R. Corman, and Jim J. Elser. 2013. *Phosphorus, Food, and Our Future*. Oxford: Oxford University Press.

Zwick, David, Marcy Benstock, and Ralph Nader. 1971. *Water Wasteland: Ralph Nader's Study Group Report on Water Pollution*. New York: Grossman.

# INDEX